Buyer Beware:
Evaluating Environmental Liabilities in Oil & Gas Transactions

Buyer Beware:
Evaluating Environmental Liabilities in Oil & Gas Transactions

Geo. L. Robertson, Editor

PennWell Books

PENNWELL PUBLISHING COMPANY
TULSA, OKLAHOMA

PennWell Publishing Company
1421 South Sheridan/P.O. Box 1260
Tulsa, Oklahoma 74101

Library of Congress Cataloging-in-Publication Data

Buyer beware : evaluating environmental liabilities in oil and gas transactions / by Geo. L. Robertson, editor.
p. cm.
Includes index.
ISBN 0-87814-453-6
1. Petroleum industry and trade--United States--Costs. 2. Gas industry--United States--Costs. 3. Liability for environmental damages--United States. I. Robertson, Geo. L. (George L.)
HD9565.B89 1995
338.2'328'0973—dc20 95-37191
CIP

Printed in the United States of America

1 2 3 4 5 99 98 97 96 95

TABLE OF CONTENTS

CONTRIBUTORS

Editor and Co-Author

Geo. L. Robertson is a practicing oil and gas and environmental law attorney in Houston, having retired as senior oil and gas attorney with Butler & Binion. He graduated from the University of Texas Law School in 1948 and has been primarily an oil and gas practitioner since that time. Mr. Robertson is Board Certified in Oil, Gas and Mineral Law by the Texas Board of Legal Specialization. He is Adjunct Professor (Energy and the Environment) at the University of Texas Law School, Austin, and at South Texas School of Law, Houston. He is a member of the American Bar Association Natural Resources, Energy and Environmental Law Section and the Oil, Gas and Mineral Law and Environmental Law Sections of the State Bar of Texas and of the Houston Bar Association. Mr. Robertson is the editor and a contributing author of *Environmental Law Simplified: A Practical Guide for Oil & Gas Operations* (PennWell Books, 1993).

Authors

Thomas A. Zabel (Chapter 5, *Field Pipeline Acquisitions*) is a partner with Burns, Wooley & Marseglia, Houston. Mr. Zabel received his B.S. in Petroleum Engineering in 1983 from Texas A & M University and his J.D., cum laude, in 1987 from South Texas College of Law. His primary area of practice is litigation, with an emphasis on oil and gas and property disputes. He is a member of the State Bar of Texas, Rocky Mountain Mineral Law Institute, and Society of Petroleum Engineers. Mr. Zabel is Board Certified in Oil, Gas and Mineral Law by the Texas Board of Legal Specialization. Mr. Zabel is a contributing author of *Environmental Law Simplified: A Practical Guide to Oil & Gas Operations* (PennWell Books, 1993).

Bernard F. Clark (Chapter 6, *Gas Plant Acquisitions*) is a partner with Butler & Binion, Houston. Mr. Clark received his B.S. in 1978 and J.D. in 1982 from the University of Texas. He is a member of the Houston Bar Association, State Bar of Texas, Rocky Mountain

Mineral Law Foundation, and Gulf Coast Cogeneration Association. Mr. Clark is Board Certified in Oil, Gas and Mineral Law by the Texas Board of Legal Specialization. Mr. Clark is a contributing author of *Environmental Law Simplified: A Practical Guide to Oil & Gas Operations* (PennWell Books, 1993).

Rosemary Stein (Chapter 8, *Offshore Transactions*) is an attorney with Exxon Company, U.S.A., Houston and concentrates her practice in environmental law. Ms. Stein received her B.A. from the University of Texas in 1973 and her J.D. from the University of Houston in 1977. She is licensed to practice law in Texas, New York, and Louisiana. Her area of practice includes all areas of oil and gas operations with an emphasis on the Oil Pollution Act of 1990. She has been a member of the U.S. Department of Interior Outer Continental Shelf Policy Committee Subcommittee on Implementing the OPA Financial Responsibility Requirements for Offshore Facilities.

Bruce A. Tackett (Chapter 8, *Offshore Transactions*) is head of media relations and principal spokesman for Exxon Company, U.S.A. He is a 1975 graduate of the University of Oklahoma. He was keynote speaker to the 1991 PetroSafe Environmental Conference and has published technical articles including environmental contingency planning for the Oil Pollution act of 1990 in Offshore Magazine. He has served as a technical consultant to other domestic companies and in 1990 was invited to Vietnam to advise on national environmental protection regulations and oil spill response capabilities. Mr. Tackett has extensive experience as a contingency planner and responder to platform, marine vessel, pipeline, overland transport, and petrochemical plant accidents.

Stanley E. Rauhut (Chapter 8, *Offshore Transactions* and Chapter 22, *Performing Cleanup Operations*) is a practicing attorney in Houston. Mr. Rauhut received his B.S. in 1965 and J.D. in 1968 from the University of Texas. He is licensed to practice law in Texas and Massachusetts. Mr. Rauhut has been admitted to practice before the U.S. Court of Appeals, First and Fifth Circuits, and the U.S. District Courts for the Southern District of Texas and for the District of Massachusetts. He is a member of the American Bar Association, State Bar of Texas, Houston Bar Association, and Maritime Law Association. His areas of practice include environmental, admiralty, international and corporate law.

John R. Hays, Jr. (Chapter 9, *Plugging Abandoned Wells;* Chapter 10, *Oil Spill Cleanup;* Chapter 11, *Hazardous Wastes;* and Chapter 12, *NORM*) is a practicing attorney in Austin. Mr. Hays received his B.A. in 1971 from Rice University and his J.D. in 1974 from the University of Texas. He is admitted to practice before the U.S. Court of Appeals, Fifth Circuit, and the U.S. District Court, Western District of Texas. Mr. Hays' areas of practice include state and federal energy, utility and natural gas law, including regulation, contract negotiation, and litigation concerning production, transmission, and distribution of oil, gas, electricity and water. He has served as Adjunct Professor at University of Texas Law School. Mr. Hays is a member of the American Bar Association, State Bar of Texas, and Federal Energy Bar Association.

Bruce Bodson (Chapter 14, *Endangered Species*) is an attorney, and is an Associate Environmental Scientist with Harding Lawson Associates and manages the Environmental Science Group in its Houston office. Mr. Bodson received his B.S. in Fishery Science and Natural Resource Recreation from the University of Arizona at Tucson, his M.S. in Environmental Science from the University of Texas at Dallas, and his J.D. from South Texas College of Law at Houston. His consulting practice focuses primarily on endangered species, wetlands, and National Environmental Policy Act compliance for the oil and gas industry. He is a member of the Natural Resource Energy and Environmental Law Section of the American Bar Association and the Environmental Law Section of the State Bar of Texas.

Susan G. Zachos (Chapter 15, *NPDES Permits,* and Chapter 16, *SPCC Plans*) is a Director in the Environmental Law Section of the law firm of Kelly, Hart & Hallman, P.C., Austin. Ms. Zachos received her B.S. and M.S. in geology, both with honors, from the University of Florida, and her J.D., magna cum laude, from the University of Houston in 1988. Her primary areas of practice include environmental law, with particular emphasis on oil and gas production issues, client counseling on environmental compliance and property transactions, Superfund negotiations, administrative litigation of regulatory issues, and administrative appeals of agency decisions. Prior to attending law school, Ms. Zachos was an exploration geologist with Exxon Company, U.S.A. in Oklahoma City, and with ESSO-InterAmerica (Exxon's South American affiliate). Ms. Zachos has written and spoken on a wide variety of oil and gas environmental topics. She currently serves on the Oil, Gas and Mineral Law Council of the State Bar of Texas and on the Environmental Committees of the Texas Independent Producers and Royalty Owners Association (TIPRO) and Texas Mid-Continent Oil and Gas Association (TMOGA).

Carie Goodman McKinney (Chapter 15, *NPDES Permits,* and Chapter 16, *SPCC Plans*) is an associate in the Environmental Law Section of Kelly, Hart & Hallman, P.C., Austin. Ms. McKinney received a B.S. in Civil Engineering from Texas Tech University in 1978, an M.S. in Environmental Engineering from Cornell University in 1984, and a J.D. from Cornell University Law School in 1987. She is licensed to practice law in Texas and California. Before becoming an attorney, Ms. McKinney was an engineer with EPA Region 6 in Dallas and at EPA's National Enforcement Investigations Center in Denver, Colorado. Ms. McKinney formerly was with McCutchen, Doyle, Brown & Enersen, San Francisco. Her primary areas of practice include client counseling on environmental issues in business transactions, environmental compliance, including hazardous and solid waste, wastewater discharge matters and international movement of wastes, and Superfund negotiations. She is a member of the Natural Resources, Energy and Environmental Law Section of the American Bar Association, the Environmental and Natural Resources Law Section of the State Bar of Texas and the Civil Engineering Academy of Texas Tech University. She has written and spoken on numerous environmental law topics.

Robert E. Steele (Chapter 17, *Clean Air Act*) is a Senior Engineering Specialist with a major oil and gas company in Houston. He received his B.A. in Chemical Engineering in 1971 and his Masters in Chemical Engineering in 1972 from Rice University, his M.B.A. in 1982 from the University of Houston and his J.D. in 1995 from South Texas College of Law. He is a registered engineer in the States of Texas and Colorado and a member of the Society of Petroleum Engineers and the American Institute of Chemical Engineers.

Estela Wackerbarth (Chapter 21, *Pre–Closing Investigation of Liability Potential*) is Senior Corporate Counsel for Safety, Health and Environmental Protection at Atlantic Richfield, Dallas. She received her J.D., with Honors, from the University of Texas School of Law and her Ph.D. (Botany) from the University of Texas at Austin. She is a member of the American Bar Association (Natural Resources, Energy and Environmental Law Section and Science and Technology Section), the State Bar of Texas, and the Dallas Bar Association (Environmental Law Section (*formerly* Chair). She is a member of the Environmental Law Subcommittee of the Texas Mid-Continent Oil and Gas Association (*formerly* Chair), and member of American Petroleum Institute working groups on various issues concerning exploration and production activities. She serves on the Board of Directors of The Western Network, an environmental alternative dispute resolution consulting firm. Prior to joining ARCO, she served as Legal Counsel at Conoco and as Assistant Regional Counsel at U.S. Environmental Protection Agency Region VI.

J. Mark Lawless (Chapter 24, *Insurance*) is a practicing attorney in Austin, primarily in environmental litigation, with an emphasis on environmental insurance disputes. He represents clients before state and federal regulatory agencies and the courts. Mr. Lawless formerly practiced in the Austin office of the Washington, D.C. law firm of Heron, Burchette, Ruckert & Rothwell. A graduate of Baylor University and the Baylor University School of Law, Mr. Lawless was admitted to practice in 1980 in the State of Texas. He is a member of the American Bar Association, the State Bar of Texas, and the Travis County Bar Association.

Gary G. Beck (Chapter 24, *Insurance*) is an insurance consultant in Georgetown, Texas. Mr. Beck is a graduate of the University of Texas and is Board Certified in Insurance Forensics. He has lectured and taught Continuing Legal Education courses for the State Bar of Texas and is a guest lecturer in Trial Advocacy/Expert Witness Examination at the University of Texas Law School.

Jurisdictional Commentators

Edward S. Renwick (*California*) is a partner with Hanna and Morton, Los Angeles. Mr. Renwick received his A.B. in 1956 and LL.B. in 1958 from Stanford University. His practice includes environmental law, and oil and gas law. He is admitted to practice before the U.S. Supreme Court, the U.S. Court of Appeals for the

Ninth Circuit, and the U.S. District Courts for the Southern District of California, the Central District of California, the Northern District of California, and the Eastern District of California. He is a member of the American Bar Association (Sections on Natural Resources, Energy and Environmental Law [*formerly* Chairman], Litigation, Antitrust Law, and Business Law); and the State Bar of California, Los Angeles County Bar Association (*formerly* Chairman of the Natural Resources Law Section), and the Chancery Club of Los Angeles. Mr. Renwick is a Fellow of the American College of Trial Lawyers. He is also a member of the Legal Committee of the California delegation to the Interstate Oil Compact Commission, and the Association of Attorney-Mediators.

Craig Wyman (*Louisiana*) is a partner with Liskow & Lewis, New Orleans. Mr. Wyman received his B.A. in 1977 from the University of Colorado and his J.D. in 1980 from Louisiana State University. He has served in the U.S. Department of Justice (Energy Section of the Antitrust Division) and as Counsel to the Louisiana Commissioner of Conservation. He is a member of the American Bar Association, Louisiana Bar Association (Council Member, Environmental Law Section), and New Orleans Bar Association.

Patricia G. Parrish (*Mid–Continent Jurisdictions: Oklahoma and Kansas*) is a partner with Musser & Bunch, Oklahoma City. Ms. Parrish received her B.A. in 1979 from Oklahoma State University and J.D. in 1982 from the University of Oklahoma. She is admitted to practice before the U.S. District Court, Western District of Oklahoma. Her principal areas of practice include oil and gas law, civil litigation, real estate law and environmental law. She is a member of the Oklahoma Bar Association (Natural Resources, Energy and Environmental Law and Alternative Dispute Resolution Sections) and the Oklahoma County Bar Association.

Phillip R. Clark (*Rocky Mountain Jurisdictions: Colorado, Wyoming, Nebraska, North Dakota, Utah, and New Mexico*) is a member in the Denver office of Holme Roberts & Owen, LLC, where his practice is primarily in the area of oil and gas and environmental law. Born in Indianapolis, Indiana, he graduated, magna cum laude, from Wabash College in 1970 and received his JD degree, cum laude, from Harvard Law School in 1976. He has written articles on producing property acquisitions and environmental matters for the Rocky Mountain Mineral Law Foundation's Special Institute on Oil and Gas Agreements, the Southwestern Legal Foundation's 37th and 41st Annual Institutes, and the University of Tulsa College of Law, has assisted in the preparation of the chapter on Competitive Leasing for the *Law of Federal Oil and Gas Leases.* He is a member of the Colorado and Denver Bar Associations, The Rocky Mountain Mineral Law Foundation, the Colorado Oil and Gas Association, the Independent Petroleum Association of Mountain States and the Rocky Mountain Oil & Gas Association. Mr. Clark served on the Colorado Oil and Gas Conservation Commission's task force on exploration and production waste regulation and bonding requirements, and recently represented COGA in a rulemaking to amend the E&P waste rules. He also currently serves on the Legal, Legislative and Regulatory Committee of the Colorado Oil and Gas Association.

PREFACE

Although this book is written by lawyers, it is addressed not to lawyers but to the oil and gas industry, and specifically to decision makers in the E&P (exploration and production) segment. The book is directed to the environmental legal concerns, that they have (or should have) upon entering into the variety of E&P transactions.

Since the book is not intended for use by legal scholars and therefore, unlike most legal writing, has a minimum of citations, nevertheless it is written with the integrity and respect for authority that characterizes a law review article or a legal treatise. On the other hand, since it is not intended as a guidebook on the details of regulation, such areas as permitting processes are not discussed.

The basic organization of the chapters is to provide a general statement of the law on the various topics (primarily federal statutes, regulations, and decisions) as well as Texas law particularly, followed by commentary by appropriate writers on laws unique to specific jurisdictions, e.g., California, Louisiana, Mid-Continent (Oklahoma and Kansas), and Rocky Mountain (Colorado, Wyoming, Nebraska, North Dakota, Utah, and New Mexico). The absence of comments for any of these jurisdictions in any particular chapter means that there are no significant local law variations from the text of the chapter.

This book does not purport to give legal advice. The reader with a particular problem should consult his own legal counsel.

ACKNOWLEDGMENTS

Putting together a book with eighteen authors was possible only by the skill, hard work, and unfailing cheerfulness and enthusiasm of our secretary, Dorothea Holloway, and the encouragement and patient support of our PennWell editors, Paula D. Wells, Sue Rhodes Sesso, Sarah Aldridge, and Don Karecki.

1

Producing Properties

In the past several years, an increasing number of U.S. domestic producing properties have come on the market, as owners have been reevaluating their properties to dispose of those that are not economic. These decisions may be made for the following reasons: (1) decline curves; (2) particular operations that are geographically isolated; (3) properties that do not fit the present policies of corporate strategy; (4) to obtain cash resources to shift operations overseas; and (5) particularly in the case of the majors, to dispose of properties with substantial reserves that are not economically recoverable if forced to bear an accounting allocation of high overhead and indirect costs. Today other reasons to dispose of particular properties include concern with either the anticipated cost of compliance with environmental regulations or with contingent liability for ghosts of years of environmental neglect. Nevertheless, there is no dearth of buyers, and competition for quality U.S. oil and gas reserves remains strong.[1]

The experienced operator who is considering a production purchase is inclined to approach a prospect in the tried-and-true way—by analyzing reservoir studies to determine the remaining recoverable reserves, which he will evaluate on the basis of his best estimate of fluctuating product prices over the remaining life of the property. He will be inclined to make a bid and pursue the purchase if the reservoir economics support the bid. He will also tend to either ignore or at least mentally discount the environmental costs of the acquisition, especially the risk of serious cleanup costs resulting from a history of environmental mismanagement in the past—perhaps far in the past. However, these costs and liability exposures must be factored into the economics of his evaluation as surely as he will factor in lifting costs. In fact, such considerations are of increasing importance in relation to the age of the production operations—that is to say, they are of especially high importance in acquisition of long-established production in old fields. It is not unreasonable to conclude in some cases that this production has been put on the market not because of depleting reserves but because the seller has made a discouraging evaluation of the cost of anticipated environmental regulation of the production, and, perhaps more importantly, the exposure to liability for cleanup.

The buyer of production must take into account a wide variety of environmental laws and regulations covering not only threatened or actual contamination of soil, groundwater, inland surface waters, coastal and marine waters, and the air, but also threats of injury to the life, health and well-being of people, animals, and plants. Of all the industries that must comply with environmental laws and regulations, clearly the oil and gas industry is one of those most heavily affected. This is particularly true of downstream operations, such as transportation, refining, and chemical manufacture. Oil and gas exploration and production, however, bear a heavy share of the burden of environmental regulation. This is the subject that is addressed in this book.

Particular environmental laws and regulations governing exploration and production that relate not only to acquisition of producing properties but also to other exploration and production transactions are discussed in later chapters. This chapter specifically addresses legal risks arising under CERCLA (also called Superfund).[2] In transactions for the acquisition of producing properties, CERCLA risks are uniquely crucial.

History and Purposes of CERCLA

Prior to CERCLA (Comprehensive Environmental Response, Compensation, and Liability Act), there was no statutory scheme for meeting one of the most egregious and widespread environmental problems in the country: the cleanup of existing toxic waste dumps. Many of these sites indeed had been long forgotten because they had been covered up and the land had been put to other uses, such as the development of residential subdivisions. The infamous Love Canal triggered congressional action, resulting in the passage of CERCLA in 1980.

The purposes of CERCLA are: (1) to identify toxic waste sites, (2) to clean them up, (3) to determine responsibility—who has to do the cleanup?—and (4) to assign liability for the cost of cleanup—who pays? Cleanup can be very expensive, and the legal liability for it can be very high—in some cases, millions and millions of dollars.

Who Is Liable?

There are four statutory categories of persons who are liable under CERCLA. These are stated in CERCLA Section 107 (a),[3] which is the heart of CERCLA. This section specifies who is liable and under what circumstances, and for what they are liable.

Loosely paraphrased, the four categories of potentially responsible parties (commonly referred to as PRPs) are:

1. Generator of hazardous substances who arranges for disposal,
2. Transporter of hazardous substances to the place of disposal,
3. Owner of the disposal site at the time of disposal, and
4. Present owner of the site.

The reason that the statute delineates four categories of PRPs is to cover situations in which the hazardous substance is generated off-site and transported to the site of disposal. In the situation where the hazardous substance is generated on the site of disposal, the only categories of PRPs that are significant are: (1) the present owner of the site and (2) the owner of the site at the time of disposal.

All of the PRPs are *jointly and severally liable.* Although this is not stipulated in the statute, it is the result of judicial decisions. CERCLA provides[4] that any defendant may seek contribution from any other defendant, either in a court action under CERCLA Section 107 (a) or in an independent action, based on an allocation by the court using such equitable factors as the court determines are appropriate. However, the EPA (Environmental Protection Agency) makes no distinction between PRPs, even though their respective quantities or toxicity of deposited hazardous substances may have been substantially disproportionate. In fact, even in judicial contribution actions, the courts have been very reluctant to make allocation of damages among co-defendants, in part because of the difficulty in proving who deposited how much and when, and exactly how toxic it was. This unrelieved joint and several liability of the defendants, of course, is an open invitation to the EPA or private party plaintiffs to press their cases against deep-pocket defendants.

> **FOR INSTANCE**
>
> Eagle Exploration owns a 40-acre oil and gas lease covering undeveloped land formerly used for a toxic substance disposal dump. Eagle disposed of no toxic substances at the dump and had purchased the lease from the Smith Estate heirs. Ace Chemical Company had a plant in the vicinity and engaged Arrow Trucking Company to get rid of its toxic waste. Arrow selected the Smith Estate land as the place to make the disposal and transported the substances to the 40 acres.
>
> PRPs include Arrow, because it selected the site and transported the hazardous substance, Ace Chemical, because it generated the hazardous substance and arranged for disposal, and the Smith Estate heirs because they presently own the land and owned it at the time of the disposal. (It is immaterial that the Smith Estate neither disposed of no toxic substances at the dump nor whether or not they consented to the disposal.)
>
> Eagle Exploration is also a PRP because it presently owns the oil and gas lease. (It is immaterial that Eagle disposed of no toxic substances at the dump.)

To simplify an analysis of CERCLA Section 107 (a), it is helpful to read a selected portion of it in the following manner: "The [present] *owner* [or] *operator* of . . . a *facility* . . . from which there is a *release* . . . of a *hazardous substance,* shall be liable . . ."

> *The term* owner-operator *is stated in the statute in the conjunctive but the courts in very early decisions concluded that the context required reading the expression in the disjunctive (and in some opinions openly chastised Congress for its inability to use the English language). The term* owner *more or less has the meaning used in the law generally. (The special problem of whether the owner of a severed mineral interest is an* owner *under CERCLA is discussed later in this chapter. Because there is no definitive law on this question, it is assumed that the present oil and*

gas lessee, as well as his lessor, is an owner.) The term operator *is not defined in the statute, but from judicial opinions should be understood to mean simply any person who is in physical control of the premises. (The term* operator *as used in CERCLA, however, has little or no relation to* operator *as it is commonly used in oil and gas exploration and production.) Thus, whether or not a conventional lessee is an owner in the common law sense, in all events he may be liable as an operator because he is in control of the premises. Also, a corporate subsidiary that is conducting operations is liable, even though its parent may be the legal owner of the premises. By the same token, a corporate officer may not be the owner of the premises, but he can be held individually liable as an operator.*

The term facility *is defined in lengthy detail in CERCLA Section 101 (9).*[5] *The remainder of the detailed definition is of no great moment in view of the concluding language that the term* facility *means* ". . . any site or area *where a hazardous substance has . . .* otherwise come to be located. . . ." *This quickly and thoroughly moots any arguments about the definition of a facility—if a hazardous substance is located there, it is indeed a facility. Contrary to common understanding, CERCLA does not use or define the terms* land *or* property. *This is important. The entirety of Eagle Exploration's 40-acre lease is not subject to CERCLA liability, only the portion of it that contains the hazardous substance. Thus, if at the time of acquisition Eagle can exclude from the lease the portion of the 40 acres where the hazardous substance is found, it may acquire a lease on the remainder of the tract without exposure to CERCLA liability. The same reasoning applies even if Eagle's lease contains 400 acres.*

The term disposal *is a key term in applying the provisions of CERCLA Section 107 (a) in all categories of PRPs other than the present owner. Disposal is defined in CERCLA by reference to another statute, the Solid Waste Disposal Act (SWDA) Section 1004 (3).*[6] *The statutory definition of disposal is all-encompassing, and it is sufficient to observe that if the hazardous waste is situated so that it* may *enter the environment, then a disposal has occurred.*

The term release *is very important. Contrary to much general understanding, the gravamen of a complaint under CERCLA is not the disposal of a hazardous substance* onto *a facility, but rather the release of the hazardous substance* from *the facility. The statutory definition of release also is all-encompassing.*[7] *Note, however, that a hazardous substance that is contained or controlled has* not *been released into the environment. Thus, it is not a release into the environment to hold a hazardous substance in inventory to use it in process of production, or to sell a product that includes a hazardous substance.*

FOR INSTANCE

Lead is a hazardous substance. But lead in inventory for manufacturing lead-based paint, using the lead in the manufacturing process, sale of lead-based paint, or applying the paint to the side of a stock tank are not releases into the environment. In all these cases, the substance is controlled or contained. However, throwing the paint cans containing residue lead-based paint into a pond is a release into the environment.

The term hazardous substance *is a controlling term because if a substance is not hazardous, it does not give rise to CERCLA liability. Thus, brick rubble, broken concrete, untreated lumber scraps, and uncontaminated soil may make for an unsightly disposal site but they do not result in CERCLA liability. BUT the statutory definition of a hazardous substance in CERCLA is very broad.*[8] *It includes:*

1. *Substances specifically identified as hazardous in EPA regulations issued under CERCLA itself,*
2. *Substances designated as toxic pollutants under the Clean Water Act,*
3. *Hazardous wastes designated under RCRA,*
4. *Hazardous air pollutants under the Clean Air Act and,*
5. *Imminently hazardous chemicals under the Toxic Substances Control Act.*

The CERCLA Petroleum Exclusion refers to that portion of CERCLA Section 101 (14)[9] (defining hazardous substance) that expressly provides that the term "hazardous substance" does *not* include petroleum or natural gas, natural gas liquids, or liquified natural gas. Unfortunately, this provision has given a false sense of security to people in oil and gas exploration and production, who erroneously conclude that because of the CERCLA petroleum exclusion, CERCLA simply does not apply to oil exploration and production. This is a false assumption because on every oil and gas lease in every day of even normal operations, the operator stores and uses hazardous substances that are not petroleum. Such products include methanol, caustic soda, zinc bromide, chlorine, sulfuric acid, ammonia, lead found in pipe dope and painting residues, and many mud additives (including, in what are now outdated drilling practices, asbestos). And if such products become mixed with crude oil, the entire mixture is a hazardous substance.

In concluding a discussion of who is liable under CERCLA, it is vital to observe what CERCLA Section 107 (a) does *not* say. It does *not* say that an owner who *intentionally* releases a hazardous substance is liable *nor* does it say that an owner who *negligently* releases a hazardous substance is liable. CERCLA liability is *liability without fault.* This is a radical departure from common law theory. The common law does not recognize liability without fault except in the limited circumstances of the old English case of Rylands v. Fletcher, which ascribes liability without fault to an owner who conducts extra-hazardous activity on his land. In Texas, under Turner v. Big Lake Oil Company, the Supreme Court has refused to apply the doctrine of Rylands v. Fletcher. There is, of course, no environmental common law and therefore common law tort theory is immaterial. The environmental law is what the statute says it is, the statute is what Congress has adopted, and what Congress adopts reflects the will of the body politic: that environmental degradation is a monumental public problem requiring extraordinary solutions. This sometimes comes as a shock to oil and gas lawyers, and to their clients as well.

Enforcement

Enforcement of CERCLA is vested in the EPA, which has very broad powers to *identify sites* that must be cleaned up, to *identify the parties* who are responsible for the cleanup (PRPs), and to *force cleanup operations.* In identifying sites, the EPA has assigned the most serious cases to its National Priorities List (NPL), which includes thousands of sites. The EPA also maintains a running list of thousands of additional sites that are under active consideration for being added to the NPL.

The EPA is not alone in identifying PRPs. It is common legal defense strategy for any PRP to identify other PRPs and to bring them into the action by virtue of the unrelieved joint and several liability ramifications of CERCLA. This burgeoning process guarantees, of course, that in any serious CERCLA action there will be a vast multiplicity of defendants. The obvious result is extremely high legal fees and many procedural delays in the prosecution of a CERCLA action.

To force cleanup operations, the EPA can issue administrative orders and, in emergency cases under CERCLA Section 106,[10] the EPA can go to court for mandatory injunctive relief. The EPA itself can clean up the site by its own contractors and then sue the responsible parties to recover the cost. Additionally, if the costs cannot be recovered, for instance, if the responsible parties cannot be found, or if they are bankrupt, or if they otherwise cannot respond to a judgment for money damages, the EPA can obtain reimbursement of its costs from the Superfund.

The EPA's administrative order (in addition to carrying the sanction of administrative penalties) or even a threat of such an order is often sufficient to obtain cleanup by a PRP. This is due to the historical fact that the cost of voluntary cleanup is quite often substantially less than that of cleanup by EPA contractors.

Enforcement of CERCLA can also be obtained by a private party who performs the cleanup and then sues the responsible parties for the costs—for instance, a buyer of a lease from a seller who is the responsible party. This can be a trap, however, for the buyer of production who reasons that he need not investigate the environmental condition of the lease he is acquiring on the theory that, if the EPA does not act, he can clean up the land and bring a private action against the responsible parties to recover his costs. In general, the right of a private party to recover the cost of cleanup is not different from that of the EPA. However, aside from the fact that the private party must advance the cost of cleanup in order to recover, he also has a serious burden of proof that the cleanup is mandated by and is in accordance with the National Contingency Plan. By express provision of CERCLA, these conditions are presumed to exist in an action by the government.

As distinguished from the legal risks in contract or tort litigation, the CERCLA exposure to liability (aside from potential criminal or administrative civil sanctions) is measured only by the cost of cleanup. This is small consolation, however, to the defendant who is liable or who is potentially liable, because the cost of cleanup can be astronomical (in an unusual case, as much as or more than the market value of the property in question).

The statement that liability exposure under CERCLA is limited to cleanup cost must be qualified by pointing out that CERCLA also expressly provides for monetary recovery by designated state and federal trustees for injury to public natural resources. This is a novel legal theory of damages, and it is incorporated into the Oil Pollution Act of 1990 (OPA 90),[11] as well as into CERCLA. It envisions monetary recovery for the right of the general public to use the damaged national resource (whether for economic or recreational benefit) or simply to enjoy the damaged national resources. The exact application of this measure of damages is currently in a stage of formulation and evolution, and to date there is little case-law development. This measure of damages, however, was seriously and concretely involved in the protracted Exxon Valdez oil spill litigation.

Third Party Defense

By definition, a PRP is only *potentially* liable. Although defenses exist, they are few in number and difficult to prove. The burden of proof is on the PRP.

The three statutory defenses stated in CERCLA Section 107 (b)[12] are: act of God, act of war, and the third party defense. The first two are unimportant as a practical matter, and the one most frequently relied upon is the third party defense. It arises out of the philosophy of CERCLA that "the polluter pays" and thus that the one who is not the polluter does not pay. Therefore, the statute provides a defense on the grounds that the pollution was the act of a third party:

- *if* the third party was not an agent or employee of the defendant,
- *and if* having discovered the existence of the hazardous substance, the defendant exercised due care as to the substance (proved by a preponderance of the evidence, in the words of the statute),

- *and if* before the act of the third party occurred, the defendant took precautions against the foreseeable acts of the third party.

> **FOR INSTANCE**
>
> Eagle Exploration is a PRP simply because it is the owner of a lease on 40 acres from which the release of hazardous substance occurred. Eagle did not place the hazardous substance on the land. It was dumped there by the driver of a truck for Arrow Trucking. Eagle proves by a preponderance of the evidence that the truck driver was not one of its employees and that Arrow Trucking was not an agent of Eagle. Eagle had fenced off the area, posted signs, and put a large sign in the middle of the only dirt road into the tract saying "NO DUMPING" (at least the jury found this was "precautions against foreseeable acts"). So far, Eagle has successfully established the third party defense.

BUT there is one more condition to the third party defense—*if* the third party was not in a contractual relation with the defendant, directly or indirectly. The exact application of "contractual relation" is not altogether clear under the cases, but CERCLA Section 101 (35)[13] says explicitly that the term "contractual relationship," for the purpose of CERCLA Section 107 (b) (3) (this is the third party defense provision discussed above) includes land contracts, deeds, or other *instruments transferring title or possession.* This is generally interpreted to mean that the defendant cannot set up the third party defense if the third party who caused the pollution as a prior owner of the land is in privity of title with the defendant. Privity of title means that there is a continuous chain of title from the defendant back to the prior owner who caused the pollution. So if any prior owner of the land caused the contamination, the present owner who is a PRP, simply because he is the present owner, cannot set up a third party defense to liability.

Innocent Landowner Defense

BUT EAGLE HAS NOT LOST THE CASE YET! The same section of CERCLA allows the defendant to prove he is an "innocent landowner," a variant of the third party defense, if the defendant can show that:

(1) Defendant acquired the facility after the disposal of the hazardous substance on the facility AND

(2) Defendant did not know that any hazardous substance was on the facility (as established by a preponderance of the evidence) AND

(3) Defendant had no reason to know that any hazardous substance was on the facility (as established by a preponderance of the evidence).

This section of CERCLA goes on to say that the defendant "had no reason to know" IF:

(a) At the time of acquisition,
(b) Defendant undertook all appropriate inquiry
(c) Into previous ownership and uses
(d) Consistent with good commercial or customary practice.

CERCLA further says that in assessing the above factors, a court shall take into account the following factors:

- Any specialized knowledge or experience of defendant
- Relationship of the purchase price to the value if uncontaminated
- Commonly known or reasonably ascertainable information about the property
- Obvious presence of contamination
- Ability to detect contamination by inspection

Pre-closing investigation of CERCLA liability potential is discussed in Chapter 21.

FOR INSTANCE

When Eagle Exploration acquired its 40-acre lease from the Smith Estate, it had no knowledge of the contamination, which occurred prior to the acquisition. Since Eagle was in

(continued)

the business of trading in oil and gas leases and seldom conducted operations, it had no knowledge or experience of hazardous substances. It paid full market value to the Smith Estate for the lease. Unfortunately, Eagle made no environmental investigation as to prior ownership or use of the property. Not only was the use of the land for a toxic dump site in prior years commonly known by people in the vicinity, but it was also obvious from even a cursory inspection on the ground that extensive dumping of wastes had taken place. In litigation of the innocent land owner issue, the court found that although it was not bound by the factors listed in CERCLA Sec. 101(35), nevertheless it was clear that Eagle Exploration had not made "all appropriate inquiry into the previous ownership and uses of the property" in accord with good commercial practice.

Severed Mineral Estates

Whether severance of the mineral estate from the basic fee title to land so as to create a separate mineral estate and a separate "surface estate" has resulted from the grant of an oil and gas lease or from a grant or reservation of the mineral estate as such, there is a legal question as to whether the owner of the severed mineral estate is subject to CERCLA Section 107 (a) liability. The question is especially serious for the owner of the oil and gas leasehold estate, which, of course, is a severed mineral estate.

The question is perplexing because CERCLA Section 107 (a), which fixes CERCLA liability, is not cast in terms of land or property as such and does not make any reference to severed mineral estates. Furthermore, there is no decided case law in any jurisdiction that addresses the question. Any legal answer, consequently, must be based solely on reasoning from the exact language of CERCLA Section 107 (a), without benefit of express statutory language and decided judicial precedent.

An argument can be made that since severed mineral estates (including oil and gas leases) are not mentioned in CERCLA, therefore CERCLA simply does not apply. This is not a reliable argument for several reasons.

First, the categories of persons who are liable under CERCLA 107 (a) include the present owner (or operator) of a facility from which there is a release of a hazardous substance. The term "facility" is not defined in the statute as being "land" or "property." However, it is defined in CERCLA[14] in an all encompassing manner, concluding with ". . . any site or area where a hazardous substance has . . . otherwise come to be located. . . ." It follows inescapably that ownership of a facility means ownership of any portion of a tract of land where the hazardous substance is located.

Second, a federal court, particularly one dealing with a situation arising in a jurisdiction without well-developed oil and gas jurisprudence, will not be inclined to override the environmental policy declared by CERCLA Section 107 (a) nor to apply (or even to understand) the nuances of oil and gas law relating to severed mineral ownership.

Third, there is the disturbing legal fact that the owner of the oil and gas mineral leasehold estate has very specific rights to use the surface of the land in the manner set forth in the lease contract, and that in most jurisdictions the severed mineral estate to a greater or less extent legally is the dominant estate over the surface estate.

The only conservative answer must be that the owner of a severed mineral estate (including the oil and gas lessee) is subject to CERCLA 107 (a) liability to the same extent as the owner of the basic fee estate. This is simply a legal risk that must be faced by the buyer of producing properties. Whether to accept the legal risk is a matter of business judgment.

If an oil and gas lease includes any site or area where a hazardous substance has come to be located, the lessee (along with various other parties who are PRPs under CERCLA Section 107 (a)) is liable for cleanup of the premises. This conclusion is the same whether the hazardous substance was deposited by the lessee himself, by the owner of the surface estate, by a prior lessee in the leasehold chain of title, or by a prior lessee in a leasehold chain of title under a prior lease that has now expired.[15]

As with any other PRP, the oil and gas lessee is entitled to assert the third party defense, as previously discussed in this chapter. As a practical matter, though, establishing the third party defense is somewhat uncertain and always depends on the existence of a number of limiting factors.

A response commonly found in contracts for the purchase of producing properties is to assert that prior to closing and after the buyer has made an environmental assessment, he either unilaterally or by agreement with the seller may partially rescind the contract as to particular properties or portions of them determined to be potential sources of unacceptable CERCLA exposure. This is not a safe response, however. It is established common law that a contract for the sale of land creates at its inception a conveyance of *equitable title,* and it is inescapable that the owner of equitable title is an owner within the meaning of CERCLA.

Thus, the only safe procedure for the buyer of producing properties is to determine prior to entering into a definitive contract for purchase whether there are areas or sites to be excluded from the transaction because of unacceptable CERCLA liability exposure. In a true letter of intent, which contains no provisions presently binding upon either party, it is acceptable to provide that the prospective buyer may have access to the premises to investigate environmental liability exposures and to exclude unacceptable areas or sites from the definitive contract.

Seriousness of CERCLA Liability

- CERCLA embodies concepts of liability that are either unique to the common law of torts or greatly expand common law tort liability concepts: strict liability (liability without fault); joint and several liability without apportionment among joint tort-feasors; retroactive liability for acts predating the statute.
- CERCLA liability can be very expensive.
- CERCLA as a statute is broad and comprehensive—there are no loopholes.
- CERCLA liability is aggressively enforced by EPA.
- CERCLA liability is aggressively enforced by the federal courts, often on the express grounds of public policy.

As stated dramatically by the Ninth Circuit Court of Appeals in a recent opinion: "CERCLA liability has been described as a 'black hole that indiscriminately devours all who come near it'."[16]

California

California has enacted its own version of CERCLA. Officially known as the Carpenter-Presley-Tanner Hazardous Substances Account Act, it is commonly referred to as "California Superfund."[17] Under

California Superfund, the state or any county of the state may order cleanup of a site that is discharging hazardous substances. The agency of the state vested with this authority is the DTSC, Department of Toxic Substances Control of the California Environmental Protection Agency. The various local counties generally enforce California Superfund through the County Health Department and through the County District Attorney.

Cleanup orders under California Superfund are not issued solely on the basis of ownership of property. Instead DTSC is required to attempt to determine who actually caused the contamination. Therefore, as in CERCLA, a former landowner or former lessee of contaminated property can be ordered to remediate the property.[18] If the designated responsible party does not do so, the DTSC is authorized to do the cleanup itself and then bring suit to recover the costs of cleanup.[19]

California Superfund has a broad definition of the term "hazardous substance" and, like CERCLA, contains a petroleum exclusion.[20] Until recently, oil field operators could feel reasonably confident that they were protected by the petroleum exclusion. However, in 1993 a federal Ninth Circuit case held that the petroleum exclusion of CERCLA did not apply to oil field production tank bottoms.[21] Because California Superfund defines a hazardous substance as any substance designed as hazardous or toxic under various federal statutes, including CERCLA, arguably oil field production tank bottoms are not covered by California Superfund's petroleum exclusion. Nevertheless, California Superfund has not been a factor in oil field cleanups. However, it is familiar to oil and gas producers because it has been applied commonly to disposal sites that contain wastes, including oil field production wastes.

One difference between California Superfund and CERCLA is that California Superfund is not retroactive. It does not purport to impose liability for acts occurring on or before January 1, 1982 if those acts were not in violation of the existing state and federal laws at that time.[22] Nevertheless, DTSC has tried to circumvent this by arguing that uncontained hazardous substances that are still leaching through the soil are still being "disposed of" until they are contained, neutralized, or removed. Under this theory, DTSC commonly orders remediation of sites containing hazardous substances even where the wastes were originally deposited well before 1982 and whether or not the hazardous substances were originally disposed of lawfully. We know of no published California opinions on this point.

Another difference between California Superfund and CERCLA is that unlike CERCLA, California Superfund does not require that releases of hazardous materials be reported to particular government agencies. There are, however, numerous such reporting requirements elsewhere in California law. It is beyond the scope of this book to list all such reporting requirements, but any oil and gas operator should make certain that appropriate reporting procedures are in place.

One feature of California Superfund and CERCLA that is of significant importance to oil and gas operators is the requirement that sellers and lessees must give notice to their purchaser or lessors of the presence of any known releases of hazardous materials on the property.[23]

Private parties who have been required to incur cleanup costs by the State of California under California Superfund or under CERCLA may seek contribution from other responsible parties.[24] It has been unclear, however, whether a party can clean up voluntarily and recover under California Superfund. Recent changes in the law have caused some commentators to conclude that the answer is yes.[25] Others conclude that the existence of a private right of action in the absence of action by the State of California is doubtful.

Acquiring oil and gas properties can also create exposure to environmental liabilities under various California statutory and common law theories. They are discussed under the California comments to Chapter 10.

Mid-Continent

In Oklahoma and Kansas, state regulatory agencies have primary regulatory authority over many aspects of the oil and gas industry, including environmental regulation. The buyer must insure that the properties are in compliance with all of a state agency's rules and regulations. Additionally, in Oklahoma and Kansas state statutory law and tort common law are relied upon more often in oilfield pollution cases than are federal laws such as CERCLA. Claims of nuisance, unjust enrichment, negligence, and trespass are often asserted in pollution litigation. Nuisance is one of the most frequently asserted common law tort claims in oilfield pollution litigation. The measure of damages in a nuisance action for injuries to land may be the reasonable costs of restoring the land. However, the courts have held that if the cost of restoring the land is greater than the diminution in market value of the land, the latter is the true measure of damages. In addition, the plaintiff in a nuisance action may recover damages for personal annoyance, inconvenience, and discomfort. The defendant in a nuisance action may be subject to injunctive order to abate the nuisance, and a buyer may be required to abate a nuisance existing at the time of his acquisition.

Additionally, the owner of the property may be required by a state agency or state law, such as the Kansas Hazardous Waste Law, to cleanup the contaminated property. This enforcement is separate from any private action for damages.

Rocky Mountain

Most of the Rocky Mountain states do not have state superfund statutes that would affect the purchaser's environmental liability in acquiring producing properties. North Dakota and Utah, however, do have superfund statutes that could have limited impact. Under the North Dakota statute concerning recovery of environmental emergency costs,[26] the state may recover from responsible parties the costs involved in the assessment, remediation, and monitoring of any environmental emergency resulting from a violation of the state's environmental statutes. Funds from the environmental quality restoration fund established by this statute may be used by the State Department of Health and Consolidated Laboratories on a case-by-case basis. Because this statute only applies to emergencies and does not appear to follow CERCLA's strict liability scheme, it should have narrow applicability.

Utah's Hazardous Substances Mitigation Act[27] establishes a limited state superfund program to undertake cleanup of hazardous waste sites, but the statute is significantly more limited than CERCLA and thus should not have significant impact on acquisitions of oil and gas properties. Except for emergencies, it is only applicable to facilities that cannot be cleaned up under other statutes, and the state must attempt to reach agreements with PRPs before it can use enforcement powers. Moreover, it is only applicable to sites either on the NPL or that have been scored by EPA for listing, and it provides for proportionate rather than joint and several liability.

In addition to these statutes, Colorado's Oil and Gas Conservation Act provides that responsible parties may be required to mitigate adverse environmental impacts from oil and gas operations and establishes an Environmental Response Fund to pay for mitigation if a responsible party cannot be identified.[28]

CHAPTER 1 ENDNOTES

1. Leslie Haynes, Guts, Feathers & All, Oil & Gas Investor, Nov. 1994, at 11 ("Buying, selling and trading [producing oil and gas] assets are standard operating procedures today, not just one-time events.").
2. CERCLA §§ 101–405; 42 USC §§ 9601–9675.
3. 42 USC § 9607(a). The full text of CERCLA § 107(a) is shown in Appendix 2–1.
4. CERCLA § 113(f)(1); 42 USC § 9613(f)(1).
5. CERCLA § 42 USC § 9601(9). The full text of CERCLA § 101(9) is shown in Appendix 2–2.
6. CERCLA § 42 USC § 6903(3). The full text of SWDA § 1004(3) is shown in Appendix 2–3.
7. CERCLA § 101(22); 42 USC § 9601(22). The full text of the pertinent part of CERCLA § 101(22) is set out in Appendix 2–4.
8. CERCLA § 101(14); 42 USC § 9601(14).
9. CERCLA § 101(14); 42 USC § 9601(14).
10. CERCLA § 106; 42 USC § 9606.
11. OPA §§ 1002(2)(A), 1006; 33 USC §§ 2702(2)(A), 2706.
12. CERCLA § 107(b); 42 USC § 9607(b). The full text of CERCLA § 107(b) is shown in Appendix 2–5.
13. CERCLA § 101(35); 42 USC § 9601(35). The full text of CERCLA § 101(35) is shown in Appendix 2–6.
14. CERCLA § 101(9); 42 USC § 9601(9).
15. A sound argument can be made that a lessee is neither in privity of title or contract with owners in the leasehold chain of title to a previously expired oil and gas lease, so that the present lessee can set up the third party defense under CERCLA § 107(b)(3) without meeting the strict requirements to prove that he is an innocent landowner that CERCLA § 101(35) does not apply in this case. There is no case law precedent, however, to support this argument.
16. Long Beach United Sch. Dist. v. Godwin, 32 F.3d 1364 (9th Cir. 1994).
17. California Health & Safety Code §§ 25300, et. seq.
18. California Health & Safety Code § 25323.5.
19. California Health & Safety Code §§ 25360, et. seq.
20. California Health & Safety Code § 25315 and CERCLA § 101(14) and California Health & Safety Code § 25317.
21. Cose v. Getty Oil Co. (9th Cir. 1993) 4 F.3d 700, 704.
22. California Health & Safety Code § 25366(a)
23. California Health & Safety Code § 25359.7.
24. California Health & Safety Code § 25363(e).
25. California Health & Safety Code § 25363(e).
26. N.D. Cent. Code ch. 23-31.
27. Utah Code Ann. §§ 19–6–301 to 325.
28. Colo. Rev. Stat. § 34–60–124.

2

Undeveloped Acreage

The acquisition of undeveloped acreage, whether by lease from the landowner or by assignment or farmout from the owner of the leasehold estate, involves essentially the same environmental risks as acquisition of producing properties. However, as described in more detail below, the lessee or buyer of undeveloped acreage has a narrower range of environmental concerns than does the buyer of producing properties. This is true, however, only when the acquisition of undeveloped acreage is not in contemplation of development.

Potential CERCLA (Comprehensive Environmental Response, Compensation, and Liability Act) liability, however, is of equal concern in the acquisition of undeveloped acreage as it is in the acquisition of producing properties. The discussion of CERCLA in Chapter 1 will not be repeated here other than to reiterate the importance of observations about who is liable, CERCLA's definition of hazardous substances and the petroleum exclusion, CERCLA's third party defense (including the innocent landowner defense), and the seriousness of CERCLA liability exposure. All such observations must be considered not only in the acquisition of producing properties but also in the acquisition of undeveloped acreage.

A technique unique to leasing acreage from the landowner is sometimes proposed for avoiding CERCLA liability. It is to restructure the provisions of the oil and gas lease itself. One such proposal is simply contractually to restrict operations to a designated portion of the leased premises. Another such proposal is to restrict the estate granted by the oil and gas lease to the mineral estate only. Both proposals negate the lessee's right to use of the surface in anticipation of development by directional drilling or by including in a unit a specific tract that will not be used as the drill site. These responses, however, are based on fallacious reasoning. CERCLA does not specify that liability arises by virtue of being the owner of or operator on the *surface* estate, but by being the owner of the *facility*—the site or area where the hazardous substance has come to be located. As discussed in Chapter 1, in the absence of any legal precedent to the contrary, it must be assumed for the purposes of CERCLA that the owner of the severed mineral estate, including an oil and gas lessee, is one of the owners of the facility.

Another proposed response is to provide for subsequent partial release of the lease of portions not used for operations (or not anticipated for use in operations), and also perhaps of subsurface depths not anticipated to be productive, in an effort to avoid CERCLA liability in respect to the part of the lease that has been released. Whether or not this response will protect the lessee from CERCLA liability arising out of acts occurring after the partial release, the lessee may be left with potential liability arising prior to the time of the partial release, since during that time he was the *owner*.

Aside from potential CERCLA liability (and also from possible liability for unplugged wells, as described in detail in Chapter 9) the buyer who does not presently consider development (for instance, the buyer who is simply building lease inventory) has no immediate environmental liability concerns. Prior to commencing development at a later time, however, the buyer must take into account such environmental issues as: hazardous substances brought onto the lease in the course of operations, spill prevention and control safeguards, handling of production wastes, possible NORM (Naturally Occurring Radioactive Material) problems, control of produced water, control of wastewater and stormwater runoff, and other environmental regulation of normal oil field operations. In addition, he must consider wetlands issues and endangered species issues that may arise later, depending on the exact location of development on the lease. All of these issues are discussed in detail in subsequent chapters of this book.

3

Joint Operations and Pooling

When a lessee enters into a joint operation with another party, he brings his lease into it with all pre-existing environmental liability exposures. If the lessee assigns an interest in the lease to the other party in order to set up the joint operation, the assignee becomes an owner for all the purposes of CERCLA (Comprehensive Environmental Response, Compensation, and Liability Act) liability. Furthermore, the assignee must take into account the problems discussed in Chapter 1 (if the lease is a presently producing property) or the more limited problems discussed in Chapter 2 (if the lease is undeveloped). If there is a cross-assignment of interests in the leases contributed to the joint operation by all parties, the result is the same for each cross-assignee.

If, however, two or more lessees contribute their leases to the joint operation but do not cross-assign interests, none of them becomes liable for pre-existing environmental liabilities burdening a lease contributed by any other joint operator, simply by virtue of entering into the joint operation.

But all of the joint operators will be at risk for CERCLA liability for any release of hazardous substances liability that arises in the course of conduct of the joint operation. And, as described in Chapter 1, since the CERCLA liability by law is joint and several, each joint operator will be exposed to full liability without reduction to his proportionate interest in the joint operation. This result is not changed by the fact that, under the joint operating agreement, one of the joint operators is appointed as Operator and the others are placed in the legal role of Non-Operators. For the purposes of CERCLA liability, each of the Non-Operators is the owner of the lease he contributed to the joint operation and notwithstanding the fact that a CERCLA hazardous substance release may have occurred solely as the act of the Operator, the Non-Operator may not assert the third party defense because the Operator is related to the Non-Operator by contract. Furthermore, CERCLA liability is imposed by statute on the owner or *operator.* Although there is no specific judicial decision on the question, by virtue of the consent or non-consent powers vested in the Non-Operator under a conventional operating agreement, almost certainly a Non-Operator will be held to have sufficient control over the property to be considered an "operator" as defined by CERCLA.

Aside from CERCLA liability for release of hazardous substances into the environment, the Operator himself will be liable for violations of all other environmental statutes that arise in the course of the joint operation, since he is the acting party. Whether a Non-Operator can escape liability on the grounds of being merely a passive party is an open question. The most analogous situation in which there is decided case law involves the question of derivative liability of officers or shareholders of a corporation for the environmental liability arising out of acts of the corporation. Under long established case law, a shareholder or officer is not personally liable for the torts of the corporation unless, under special circumstances, the court "pierces the corporate veil" and finds such personal liability. As discussed in greater detail in Chapter 23, the clear trend of federal judicial decisions in environmental cases, however, is to ignore or override this common law doctrine. An officer or shareholder may be held liable and is not shielded by the corporation from liability for environmental offenses, especially where the officer or shareholder personally participated in the act or omission giving rising to the environmental violation. More specifically, an officer has been held personally liable for the corporation's environmental violations simply because he had corporate authority or was in a position to prevent the violation. Based on this analogy, the Non-Operator, although he is in a merely passive role, may be held liable for the environmental violations of the Operator. This is particularly true if the Non-Operator personally participated in the violation, had contractual authority under the joint operating agreement, or was in a position to prevent it.

Even so, the Non-Operator cannot rely on protection based on this analogy with any degree of certainty. If the environmental violation arises directly from an act that was authorized by the Non-Operator expressly or by clear implication, such as location of a well or other conduct of operations on wetlands as defined under the Clean Water Act or operations resulting in the unlawful taking of threatened or endangered species under the Endangered Species Act, the Non-Operator may be liable directly.

Execution of a **unit designation** under a lease pooling clause or joining with other lessees in execution of a unit designation does not of itself create CERCLA liability nor liability under any other environmental statutes. By the same token, joining in an agreement for unitized operations does not of itself create such liability.

California

Oil and gas operators should be aware of two aspects of California law that arguably could give rise to unanticipated liability in connection with joint operations and pooling. The first involves community leases and the second involves joint operating agreements.

A community lease, as that term is used in California, is a single lease covering multiple, separately owned tracts of land. Such leases are in common usage in California. Early case law in California interpreted community leases as giving rise to cross assignments of interests so that each community lessor was deemed to have transferred to his fellow community lessors a share of his mineral fee and to have received a corresponding fractional share of the mineral fee in each of the other community lessors tracts, at least for so long as the community lease was in existence.[1] Thus, if one were to buy a mineral fee in a tract that was subject to such a community lease, he may have acquired an interest in all of the other tracts committed to the same lease. If there were environmental problems attached to the operation of that lease, then conceivably, he, as well as all of the lessors, might have CERCLA liability.

There is also a possible hidden risk in signing a joint operating agreement. Oil and gas operators should be aware that by doing so, they may be taking on joint and several liability under the California partnership laws even though they are designated as non-operators and the joint operating agreement says that each joint operator is only liable for his proportionate share of any costs or liabilities. This is because the typical joint operating agreement in California provides that the various joint operators share in the profits and losses of the jointly operated properties according to certain percentage shares. Arguably, this may create a partnership as a matter of law. The California version of the Uniform Partnership Act declares that the receipt by a person of a share of the profits of a business is prima facia evidence that he is a partner in the business.[2] If he is a partner, he is jointly liable with the other partners for all of the partnership's liabilities.[3]

Mid-Continent

In a federal case arising in Oklahoma,[4] which was brought up on the common law tort theory of nuisance, it was held that a unit non-operator is liable for a nuisance created and maintained by the unit operator during the time the non-operator was an owner in the unit. This is in accordance with the Oklahoma rule that the operator is agent of the non-operator. The same result may be obtained under the Oklahoma and Kansas rule that, where there is joint ownership, a sharing of profits and mutual participation in operations, the non-operator may be held liable as a joint venturer or as a mining partner.

CHAPTER 3 ENDNOTES

1. Tanner v. Title Insurance & Trust Co. (1942) 20 Cal.2d 814, 820.
2. California Corporations Code § 15007(4).
3. The author successfully made this argument at the trial court level in California a number of years ago.
4. Branch v. Mobil Oil Corporation, 788 F.Supp. 531 (W.D. Okla. 1991).

4

Non-Operating Interests

In general, acquisition or ownership of a non-participating royalty interest, an overriding royalty interest, a production payment interest, or net profits interest does not result in exposure of the owner to liability under any environmental statutes. This is related to the fact that the ownership is entirely passive and does not involve the conduct of activities that are regulated by environmental statutes.

This statement does not include a non-operating working interest, because it is not entirely passive. Potential environmental liability exposure of non-operating working interest owners is discussed in Chapter 3. This statement also does not include ownership of a carried working interest, since in most cases the owner of a carried interest is simply a working interest owner (whether he is carried for drilling costs, or for completion costs, or for operating costs prior to payout) who by contract is not obligated to pay for those costs that are particularly specified in the instrument creating the carried interest. The carried party may be liable under environmental statutes either as an operating working interest owner or as a non-operating working interest owner, depending upon his role under the contract creating the carried interest. The fact that, as between the parties, the carrying party, and not the carried party, is liable for costs during the period of the carry, does not affect the carried party's statutory liability exposure.

Under the common law of property in some states, the owner of a non-participating property interest is the owner of an interest in land. In other states, by common law or by statute, the non-participating royalty owner is entitled to the benefits of the public records doctrine and other laws *as if* he owned an interest in land. Nevertheless, it is reasonable to conclude in either case that the non-participating royalty owner is not an "owner" for the purposes of CERCLA Section 107 (a) liability.[1]

In a recent United States Court of Appeals decision,[2] it was held that the owner of a pipeline easement across a toxic waste dump who did not participate in the deposit of substances into or release of substances from the dump was not an owner for CERCLA (Conprehensive Environmental Response, Compensation, and Liability Act) liability purposes. This decision was based on the grounds that common law courts have consistently distinguished between ownership of the mere easement and ownership of the land that is burdened with the easement. Although this case is not directly applicable, the same reasoning applies to non-participating royalty and similar non-operating interests. Even in jurisdictions that hold that a royalty interest is an interest in land, it is characterized as a non-possessory interest.

In the case cited, the court observed that the owner of an easement who conducts physical activities on the land might be liable as an "operator" under CERCLA.[3] Thus, a non-participating royalty owner who conducts activities on the land (for instance, activities in taking in kind his royalty interest in production, and storing or transporting it) might be held liable as an "operator" under CERCLA even though he is not an "owner" by definition.

California

There is some question in California as to whether owners of non-participating royalty interests, overriding royalty interests, production payment interests, and net profit interests that are entirely passive are insulated from liability predicated on ownership under CERCLA. The reason for this is that in California, the owner of such an interest is generally held to own an interest in the hydrocarbon mineral estate itself. For instance, in an early 9th Circuit case purporting to apply California law, a royalty owner was held to be not simply a creditor of a bankrupt oil and gas operator but also the owner of his royalty share of production. His royalty share of oil in production tanks and of cash in the bankrupt oil and gas operator's bank account was not available to the bankrupt's general creditors. It belonged to the royalty owner.[4] Because owner/lessors of land have been held to be responsible parties in connection with CERCLA liability resulting from their lessee's actions on their property, it is arguable that the same logic might be applied to the owners of interests in the mineral estate even though by their very nature the interests are entirely passive.

CHAPTER 4 ENDNOTES

1. See Chapter 1 for discussion of the meaning of "owner" under CERCLA § 107(a).
2. Long Beach United Sch. Dist. v. Godwin, 32 F.3d 1364 (9th Cir. 1994).
3. See Chapter 1 for discussion of the meaning of "operator" under CERCLA § 107(a).
4. Consolidated Royalties, Inc. v. Ashton (9th Cir. 1942) 132 F.2d 226, 229.

5

Field Pipeline Acquisitions

This chapter specifically addresses environmental concerns associated with field pipeline systems, commonly referred to as gathering or feeder lines. Field pipelines systems gather production directly from the site and transport it to a field processing facility or to a main transmission line. The term "field pipeline system" as used throughout this chapter refers not only to the pipeline itself, but also to the related pipeline facilities such as inlet separators, meter stations, pigging traps, pipeline drips, treating facilities and compressors, and associated real property rights.

Basically, the same environmental considerations associated with producing properties discussed in Chapter 1 apply to field pipeline systems. Specifically, CERCLA (Comprehensive Environmental Response, Compensation, and Liability Act) "owner" and "operator" liability with respect to hazardous substances applies equally to field pipeline systems as to producing properties. Thus, prior to purchasing a field pipeline system, the buyer should consider conducting an environmental due diligence review, including a site assessment of key areas along the pipeline system.

Although field pipeline systems customarily transport virgin crude oil and natural gas, both of which fall within the CERCLA petroleum exclusion and RCRA (Resource Conservation and Recovery Act) oilfield waste exemption, other chemicals, lubricants, solvents, and materials that may have been used in connection with field pipeline operations are not covered by such exclusion or exemption. Such substances may constitute hazardous substances under CERCLA or hazardous wastes under RCRA, or both, which, if improperly disposed of along the field pipeline system, could expose an owner or operator of the pipeline system to CERCLA liability.

Since the CERCLA petroleum exclusion and RCRA oilfield waste exemption are described in more detail in other chapters, they will not be further discussed in this chapter other than to briefly discuss a recent clarification action by the EPA (Environmental Protection Agency), in which it addressed the scope of the RCRA oilfield waste exemption in regard to pipelines transporting crude oil and those transporting natural gas.[1]

Prior to the clarification, uncertainty existed as to the applicability of the exemption to pipelines transporting crude oil or natural gas. Under the clarification, the EPA specified the scope of such exemption by making a distinction between field pipelines transporting crude oil and those transporting natural gas.

Crude oil that is produced from a reservoir and any related produced water are typically sent to a tank battery where the oil and water are separated naturally by the force of gravity. This separation is sometimes enhanced by the use of heat in a heater treater unit. Most water is separated from the oil at the tank battery and does not enter the crude oil pipeline system. In the case of crude oil, the EPA has determined that all production-related activities for purposes of the RCRA oilfield waste exemption occur as part of primary field operations at or near the wellhead, and that wastes that are generated as part of the process of transporting crude oil away from primary field operations are not exempt. Specifically, the EPA has determined that for crude oil production, custody transfer of the crude oil, or in the absence of a custody transfer, the end point of initial production separation of the oil and water, defines the end of the primary field operations and the beginning of transportation. For crude oil, only waste generated before the end point of primary field operations and prior to the point of transportation are exempt.

Any waste generated by the transportation or handling of crude oil after custody transfer or, in the absence of custody transfer, after the end point of initial product separation of oil and water, does not fall within the petroleum exclusion. Examples of nonexempt waste from transportation include pipeline pigging waste (unless associated with a production facility), contaminated water and soil resulting from spills, and other wastes associated with pipeline transportation.

As distinguished from crude oil, natural gas, which is produced from a reservoir through one or more wells, is typically separated from hydrocarbon liquids (condensate) and some water at the wellhead. Due to economic constraints, however, gas from several wells is generally commingled and sent to a central gas processing plant, where additional water and other impurities are removed.

Although the ownership or custody of natural gas commonly changes hands between the wellhead and the gas plant, the EPA has concluded that the removal of impurities from the gas at the gas plant is still a necessary part of the production process for natural gas. Unlike crude oil lines, the EPA has concluded that because movement of natural gas between the wellhead and the gas plant is considered a necessary part of the production operation, uniquely associated wastes derived from the production stream along the field pipeline system (i.e., produced water and gas condensate) are exempt wastes, even if title to the natural gas has changed custody between the wellhead and the gas processing plant. However, the EPA has noted that some wastes generated at the production stage may not be uniquely associated with the natural gas production stream and thus are not exempt (such as pump lube oil and waste mercury from meters and gauges). Similarly, soil contaminated by spills of wastes that are not uniquely associated with production operations, such as by mercury, waste oil solvents, or chemicals are not exempt waste. In its Clarification Action, the EPA indicated that waste generated at compressor stations and facilities located downstream from the gas plant are not covered by the EPA exemption. Presumably, wastes generated at compressor stations and facilities located upstream from the gas plant, which are uniquely associated with oil and gas exploration and production, will be considered exempt by the EPA.

Production of elemental sulfur from hydrogen sulfide gas at a leased facility or at a gas plant is considered treatment of an exempt waste since this gas is considered a uniquely associated waste. This includes iron sponge from a field treating vessel.

As with producing properties, a primary concern in the purchase of a field pipeline system is to determine whether it is contaminated by substances such as polychlorinated biphenyls (PCBs) or mercury. This determination is based on the location of the key areas where the contamination may exist. These areas and the possible contaminants associated with such areas include:

Area	Possible Contaminant
Meter station site	Mercury
Compressor site	PCBs, waste lubrication oils, cooling waters, hydraulic oils, cleaning chemicals, and used filters

(continued)

Area	Possible Contaminant
Treating facilities site	Ethylene, diethylene, triethyline glycol, alumina, silica-gel, silica-alumina beads, used filters
Sweeteners	Amine, sulfinol, iron sponge, caustic solvents, and used filters
Inlet separators	Produced water, inlet filter media, fluids from corrosion treatments, pipe scale
Pigging traps and pipeline drips	Hydrocarbons, pipe scale, sediments, waste water

Of these contaminants, PCBs in pipelines have gained the most attention. In the mid-to-late 1950s, many pipeline companies used lubricating oils that contained PCBs. These lubricating oils were used in pipeline compressors through the early 1970s. Although PCBs are encountered primarily with respect to main transmission lines, it is possible that owners or operators of field pipeline systems may have used lubricating oils containing PCBs.

Although PCBs are not considered hazardous wastes under RCRA, they fall under the jurisdiction of TSCA (Toxic Substances Control Act). PCB contamination, if it exists at all, is typically found in the area of compressor station sites or in pits that have been used for the disposal of pipeline liquids and other wastes generated by the operation of the pipeline system. Since PCB contamination is generally associated with transmission lines, it will not be further discussed here.

Many metering systems associated with pipeline operations may have utilized mercury-based flow meters or manometers monitoring flow rates. In some cases, the use of mercury-field meters has resulted in contamination of soils with elemental mercury.

Contamination of soils at metering sites can result from meter leaks or spills, pressure surges, vandalism, or other causes. Since a typical meter contains only 8 to 10 pounds of mercury, the area of contamination is usually confined to the soils within the immediate vicinity of the meters themselves. Such areas of mercury contamination usually do not exceed one or two cubic yards in size. Careful review of historical records and physical examination around current meter locations should be made.

Environmental contamination with regard to separators, dehydrators, sweeteners, and compressors is discussed in more detail in Chapter 6.

California

The acquisition of field pipelines can also create exposure to environmental liabilities under various California statutory and common law theories. They are discussed hereafter under the California comments to Chapter 10.

In addition, the owner of a field pipeline should be aware of the California Pipeline Safety Act of 1981.[2] Under this Act, the California State Fire Marshal exercises safety regulatory authority over certain hazardous liquid pipelines. The term "pipeline" probably excludes field pipelines systems because it excludes:

a. Transportation of petroleum in onshore gathering lines located in rural areas.[3] (A gathering line is a pipeline 8" or less in nominal diameter that transports petroleum from a production facility.[4] A production facility refers to the piping or equipment used in the production, extraction, recovery, lifting, stabilization, separation, or treatment of petroleum or associated storage or measurement.[5])

b. A pipeline for the transportation of crude oil that operates by gravity or at a stress level of 20% or less of the specified minimum yield strength of the pipe.[6]

c. Transportation of a hazardous liquid by a flow line.[7] (A flow line is a pipeline that transports substances from the wellhead to a treating facility or production storage facility.[8])

However, that same act requires every rupture, explosion, or fire involving a pipeline, including an exempt pipeline system as set forth above to be immediately reported by the pipeline operator to the local fire department and to the California Office of Emergency Services. In addition, the operator must, within 30 days of the rupture, explosion, or fire, file a report with the State Fire Marshal containing all information required by the State Fire Marshal.[9] The phrase "rupture" includes "unintentional" liquid leak.[10] Any willful violation of any provision of the Hazardous Pipeline Safety Act shall, upon conviction, be subject to a fine of not more than $25,000, imprisonment for a term not to exceed 5 years or both.[11]

Louisiana

In Louisiana, in addition to the waste management concerns raised by the chapter's main text, field pipeline systems located within state waters are also subject to special site clearance regulations. The Pipeline Division of the Office of Conservation administers an underwater obstructions program that includes the timely removal of field pipeline systems "associated with abandoned oil and gas facilities located in coastal waters and bays." Approval of a site clearance plan must be obtained from the agency.[12]

CHAPTER 5 ENDNOTES

1. Clarification of the Regulatory Determination for Waste From the Exploration, Development and Production of Crude Oil, Natural Gas and Geothermal Energy, 58 Fed. Reg. 15, 284 (1993) (to be codified at 40 CFR Part 261).
2. Government Code § 51010.
3. Government Code § 51010.5(a)(4).
4. Government Code § 51010.5(f).
5. Government Code § 51010.5(g).
6. Government Code § 51010.5(a)(3).
7. Government Code § 51010.5(a)(6).
8. Government Code § 51010.5(b).
9. Government Code § 51018(a).
10. Government Code § 51018(c).
11. Government Code § 51018.7.
12. LAC 43:XI.301 et seq.

6

Gas Plant Acquisitions

Aside from operations at the well for first-cut separation of liquid hydrocarbons from the gas stream, acquisition of producing properties frequently includes off-lease facilities for recovery of additional liquid hydrocarbons or for removal of water or impurities to meet gas purchase agreement requirements. Such facilities are often referred to generically as gas plants, although they vary considerably in design and function. Although an off-lease gas plant can generate environmental problems similar to those of lease operations, such problems can be different and far more serious, particularly problems arising under the Clean Air Act.

CERCLA *Liability*

Since the gas plant ordinarily is located either on a fee tract or a surface lease, the buyer becomes either an owner or an operator for the purposes of CERCLA (Comprehensive Environmental Response, Compensation, and Liability Act) liability arising from prior ownership or uses, simply because he is the present owner or operator of the site. This liability is discussed in greater detail in Chapter 1. It is therefore necessary to conduct an appropriate environmental site assessment, as in the case of a lease acquisition.

In addition to CERCLA liability for prior contamination, operation of the gas plant can also create CERCLA liability because of the generation of hazardous substances. Due to the CERCLA petroleum exclusion, spills of crude oil and other produced hydrocarbons not contaminated with other hazardous substances are not a matter of concern. However, many catalysts, solvents, cleaning solutions, along with paint and other materials utilized in gas plant operations are not covered by the CERCLA petroleum exclusion. An on-site investigation of potential releases of such particular hazardous substances should be made prior to closing and contracting if possible.

Typically, hazardous substances generated in gas processing operations are transported for disposal off-site. If the gas plant is acquired through merger or the new owner is deemed a "continuing enterprise" of the former company, the new owner may acquire not only the gas plant but also liability for its seller's waste, which may have been improperly disposed of years before the sale. Whether or not a buyer becomes liable for such waste, it is important for the buyer to assess the need to investigate the seller's past disposal practices, including any on-site investigation of such disposal facilities and the desirability of continuing the seller's hazardous substance off-site disposal arrangement. Such assessment includes not only an investigation of the company that owns and operates the disposal facility but also a determination of the identity and financial responsibility of other operators who are using the same site. Otherwise, the buyer may find himself to be the only deep-pocket defendant of the several who may be jointly and severally liable.

Gas Plant RCRA Hazardous Wastes

For the most part, wastes generated in the operation of a gas plant are covered under the RCRA (Resource Conservation and Recovery Act) oilfield waste exemption, which is discussed in detail in Chapter 11. The test of the exemption is whether the wastes are generated in primary field operations in connection with processes "uniquely associated" with the production of natural gas. For natural gas, primary field operations include those production-related activities at or near the wellhead and at the gas plant (regardless of whether or not the gas plant is at or near the wellhead) but prior to transport of the natural gas from the gas plant to market. Since the movement of natural gas between the wellhead and the gas plant is considered a necessary part of the production operation, uniquely associated wastes derived from the production stream along gas plant feeder pipelines are considered exempt wastes under RCRA.

Generally, these are waste materials that come into contact with the gas stream in the treating process on its way to the transmission pipeline (for example, filters, sieves, and produced waters). Gas plant wastes that do not ordinarily come into contact with the gas stream are not exempt (such as painting and cleaning wastes, pump lube oil, hydraulic fluids, and waste mercury from meters and gauges).

The EPA (Environmental Protection Agency) has identified and listed wastes that are exempt from regulation as hazardous wastes that are generated at gas plants by contact with the production stream during the removal of produced water or other contaminants from the produced raw natural gas. It also has identified and listed gas plant wastes that are not exempt as hazardous wastes under the RCRA. Exempt wastes include:

- Gas plant dehydration wastes, including glycol-based compounds, glycol filters, filter media, backwash, and molecular sieves;
- Gas plant sweetening wastes for sulfur removal, including amine, amine filters, amine filter media, backwash, precipitated amine sludge, iron sponge and hydrogen sulfide scrubber liquid and sludge;
- Cooling tower blowdown;
- Spent filters, filter media, and backwash (assuming the filter itself is not hazardous and the residue in it is from an exempt waste stream);
- Liquid hydrocarbons (crude oil, condensate, and NGLs) removed from the production stream but not from oil refining;
- Gases removed from the production stream, such as hydrogen sulfide and carbon dioxide, and volatilized hydrocarbons; and
- Light organics.

Non-exempt wastes include:

- Gas plant cooling tower cleaning wastes and solvents;
- Painting waste;
- Service company wastes;
- Lubrication oils;
- Contaminated soils;
- Waste compressor oil, filters, and blowdown; and
- Pesticide wastes.

In addition, any waste generated by non-production related activities described by the EPA as "manufacturing processes" associated with the gas plant operations are not exempt. These non-exempt manufacturing activities include operations that go beyond the removal of impurities from the raw gas and the physical separation of the gas into its component fractions. These manufacturing activities, generally occurring at only the larger gas plants, are similar to those in petrochemical refinery operations, such as cracking and reforming the molecular structures of the various gas fractions and the addition of odorants or other substances. The RCRA exemption ends in the gas plant once manufacturing begins or, if no manufacturing occurs, at the point at which the natural gas leaves the plant for transportation to the market.

Compression

In order to meet the operating pressure of the transmission pipeline and to move the gas through the plant, the first phase of gas plant operations may be to increase the gas pressure through compression. Since most methods of processing the raw natural gas in the plant require constant manipulation of the gas pressure as a part of the process for stripping out natural gas liquids (NGLs), multiple gas compressors are utilized in many areas of gas plant operations. Almost all compressors are powered by internal combustion engines. The most significant environmental concern is monitoring and controlling exhaust emissions from the compressor engines, which is discussed in Chapter 17. In addition, routine servicing and maintenance of the compressors can also lead to environmental liability. When serviced, waste lube oil, spent solvents, and engine coolant can contain hazardous chemicals not covered either by the petroleum exclusion under CERCLA or the oilfield waste exemption under RCRA. Since these hazardous substances may have been improperly disposed of on site, close attention to soil and surface conditions around this equipment and adjacent pits is well advised.

Metering

In addition to compression, the gas and NGLs are usually metered at both the inlet and outlet tailgate of the plant. Some older meters may have used mercury to measure gas flow. A typical mercury meter contains 8 to 10 pounds of mercury. Mercury may have leaked or spilled during servicing of such meters or due to other causes, including equipment failure, pressure surges, vandalism, and operator error. Although an inspection may not reveal any mercury meters currently in use, this does not mean that mercury meters have not been used in the past. If spilled, mercury contamination is generally limited to a small area around the meter. The volume of mercury-contaminated soils generated by the remediation of individual sites is generally small, usually about one or two cubic yards. A careful review of gas plant historical records and physical examination around current meter locations should be made.

Treating Operations

An important function of the gas plant is to treat the raw natural gas to remove remaining water and impurities that field area production and gathering operations failed to catch. Discussed below are the three main treating operations performed at gas processing plants: inlet separation, dehydration, and sweetening operations.

Inlet Separation

Raw gas and produced fluids gathered from the field are initially separated at the inlet of the gas plant. The RCRA exempt oilfield wastes typically associated with inlet separation include produced water, slop oil (crude), and condensate, as well as pigging materials, inlet filter media, fluids from corrosion treatments, and small amounts of solid material (pipe scale, rust, and reservoir formation materials). Separators are designed to collect the produced wastes in tanks or vessels. However, since older gas plants may have formerly used open earthen pits for collection, an investigation should be made to determine their location. In the event that more fluids are produced than can be handled in the primary containers, the equipment on some of the older plants is designed with relief valves, which can empty into either contained vessels or open emergency pits. Any existing emergency pits for discharges from relief valves on inlet separators as well as any former abandoned pit sites should be inspected to insure compliance with applicable regulations for open pits. In addition, any fluids diverted to the emergency pits should be recovered as soon as practical. If there is reason to believe that hazardous wastes may have been improperly disposed of in the pits along with raw gas and produced waters, it is prudent to test the pits and surrounding soils for the presence of contamination. In addition, any abandoned pits located on-site should also be similarly tested.

Produced water can be disposed of either through properly permitted Class II injection wells or by being trucked to a saltwater disposal facility. If a disposal well is located on-site, all necessary permits, inspection records, and maintenance accords should be examined to make sure they are in order. In any event, care should be taken to verify that such disposal methods comply with local regulations. Surrounding areas should be inspected for visual signs of uncontrolled surface disposal of produced waters evidenced by white, crusty soil with sparse or distressed vegetation. Although produced waters may be exempt under the RCRA oilfield waste exemption, if the waters reach navigable waters of the United States (any stream, creek bed, or lake), such improper disposal may give rise to liability under the Clean Water Act.

Dehydration

In order to further reduce the water vapor content of produced natural gas, water is brought into contact with liquid or solid desiccants. Waste generated during the dehydration process consists of ethylene, diethylene, and triethylene glycol-based fluids, glycol filters, condensed water, and solid desiccants such as alumina, silica gel, silica alumina beads, or molecular sieve. These RCRA exempt liquid and solid wastes should be either recycled, returned to the vendor, or disposed of as a nonhazardous waste at a properly permitted saltwater disposal facility or in a Class II nonhazardous waste-injection well.

Air emissions from the reboiler still vents of glycol dehydration units can include benzene, toluene, ethyl benzene and xylene isomers (BTEX). Other volatile organic compounds (VOC) include ethylene glycol, formaldehyde, and acetaldehyde. Under the Clean Air Act, glycol dehydration units that emit 10 tons per year of any of these hazardous air pollutants or 25 tons per year of any combination of such hazardous air pollutants are subject to regulation requiring installation of maximum control technology to reduce emissions. At a gas plant, the figure for the total tons per year emissions of air toxics include not only emissions from dehydration units but also fugitive emissions (such as from pipe connections, valves, and flanges), tank emissions, engine emissions, and heater emissions. Gas plants located in non-attainment areas are subject to even more stringent emission guidelines and operating permit requirements. A thorough review of local regulations and existing gas plant air permits is critical to an overall assessment of potential Clean Air Act liabilities at the gas plant.

Sweetening/Sulfur Recovery

In addition to reducing the water content, the produced gas can also require sweetening, depending on the chemical composition. Sweetening primarily consists of lowering the hydrogen sulfide and carbon dioxide content of the produced natural gas. In the industry, hydrogen sulfide is most commonly removed by amine treatment. Wastes generated in amine sweetening include spent amine, used filter media, and acid gas, which must be flared or sent to a sulfur recovery facility. If flared, emissions must comply with permitted levels. Other processes for removing hydrogen sulfide include iron sponge, caustic solutions, and other sulfur converting materials. In the iron sponge process, which is used in smaller operations, wood shavings are coated with iron oxide, which then reacts with the hydrogen sulfide, generating as waste iron sulfide and wood shavings. Since spent iron sponge is capable of spontaneous combustion, it should be disposed of in accordance with state regulatory rules. In Texas, it should be allowed to fully oxidize before being properly disposed of at a permitted disposal facility under a Railroad Commission of Texas Statewide Rule 8 minor permit. Small volumes of hydrogen sulfide can also be removed by contact with a caustic solution, typically sodium hydroxide. The primary waste from caustic treating is spent caustic solution. When higher concentrations of hydrogen sulfide gas are encountered, the hydrogen sulfide is removed using amine or sulfinol solutions and is then further treated to form elemental sulfur.

In addition, molecular sieve absorbents are used to remove hydrogen sulfide, mercaptans, and heavier sulfur compounds. Waste materials in such processes can include water vapor; regeneration gas, which is used as a fuel source or recombined with residue gas and sent to sales; and spent molecular sieve. Such sweetening wastes are included within the RCRA Oilfield Waste Exemption.

Recovery of Natural Gas Liquids

Many gas plants process gas to separate out the NGLs from the gas stream. The NGL components are initially separated as a mix (which is called "raw mix" or "NGL mix"), and can then be separated by their constituent fractions—ethane, propane, butanes, and natural gasoline.

NGL extraction can use compression processes, cooling processes, absorption processes, or cryogenic processes, or a combination of these, none of which involve the introduction of catalysts or chemical reactions. Since there is no change in the molecular structure of the various hydrocarbons during gas processing no waste by-product is generated from the product stream. The mechanical apparatus surrounding the product stream does, however, generate wastes.

Wastes generated during processing include waste waters, cooling tower water, and boiler blowdown water (including any bactericides and corrosion inhibitors dissolved in such water)—all of which are deemed uniquely associated with the production of natural gas and are therefore exempt under the RCRA Oilfield Waste Exemption. Since waste waters are continuously discharged from gas plant operations, the owner of the gas processing facility may be required to haul away produced waters and dispose of them in a Class II injection well or to obtain a National Pollutant Discharge Elimination System Permit and to operate within its limits. In addition, the facility must have an appropriate Spill Prevention Control and Countermeasure Plan (SPCC Plan), which is described in more detail in Chapter 16, and storm water discharge permits. If wastewaters are disposed of through a Class II injection well, the well must have a permit. However, some other wastes generated at this production stage may not be uniquely associated with the natural gas production stream and are, therefore, not RCRA exempt. These include compressor, generator, and pump lube oil; cooling tower cleaning wastes; painting wastes; biocides; bactericide and corrosion inhibitor wastes (not in solution with the cooling tower or boiler water); and waste mercury from meters and gauges. Any mixture of RCRA exempt wastes contaminated by non-exempt waste can result in the entire waste stream losing its RCRA exempt status and perhaps having to be handled as a hazardous waste. Similarly, soils contaminated by spills of wastes that are not uniquely associated with production operations, such as by mercury from gauges, are not RCRA exempt wastes.

TSCA *Reporting* Requirement

Section 8(a) of the Toxic Substances Control Act (TSCA) requires that anyone manufacturing, importing, or processing certain chemical substances for commercial use listed on the Master Inventory File must file a report with the EPA under the Inventory Update

Rule to be updated every four years, beginning in 1986. Liquefied petroleum gas was included in the Master Inventory File in 1985 by the American Petroleum Institute. As a result, operators of natural gas plants producing 10,000 pounds annually of ethane, propane, butane, or natural gasoline (approximately 300 gallons of any one substance) are required to file reports every four years. Industry efforts by the Gas Processors Association in 1994 to change the EPA reporting rules resulted in an informal settlement option for natural gas processors who failed to file in 1986 or 1990. The window to "opt in" to such settlement closes in 1995. Without such informal settlement, the penalty assessable for failure or refusal to establish or maintain records and submit reports required under TSCA can be a civil fine up to $25,000 per violation; each day such a violation continues is considered a separate violation. One industry newsletter reports that a major independent was fined almost $500,000 in 1993 for being more than four months late in submitting information on NGLs it produced at 18 of its gas plants as required by Section 8(a) of TSCA.

DOT *Certification*

All gas plants that offer for interstate transportation more than 3,500 gallons of crude oil, condensate, or NGLs in a single container or vessel must obtain a certificate from the Department of Transportation. A copy of the registration certificate must be kept at the company's principal place of business. Failure to register with the DOT can subject companies to civil and criminal penalties.

California

The acquisition of a gas plant can also create exposure to environmental liabilities under various California statutory and common law theories. They are further discussed under the California comments to Chapter 10.

Louisiana

This chapter's discussion of mercury meters has particular relevance to Louisiana operations and environmental site assessments associated with gas plant acquisitions. Possible contamination resulting from mercury meters has been the subject of special inquiries in recent years by the Louisiana Department of Environmental Quality.

7

Lending Transactions

Many oil and gas acquisitions are accompanied by a lending transaction. Although this book is not addressed to lenders, as such, nevertheless, a buyer who contemplates a contemporaneous lending transaction must be aware of the lender's concerns about exposure to environmental liabilities.

As a matter of good lending practice, a lender accepting oil and gas collateral must be concerned about its borrower's exposure to environmental liabilities for the same reason that the lender is concerned about the borrower's compliance with any governmental regulation of the borrower's operations. In the case of environmental regulation, this lending concern extends to such matters as prior contamination of soil, groundwater, and surface water; handling of production wastes, wastewater, and stormwater runoff; and spill prevention and control. These topics are discussed in more detail in other chapters.

The lender must be particularly concerned with real or potential violations carrying sanctions with such potentially adverse economic impact that they might seriously impair the borrower's general financial condition, as for instance, extensive cleanup obligations under CERCLA (Comprehensive Environmental Response, Compensation, and Liability Act).[1]

Even more importantly, the lender must be concerned with the lender's own direct exposure under CERCLA as an owner or operator of a facility. CERCLA contains an express provision for a secured creditor exemption,[2] which states that the term "owner" or "operator" for CERCLA purposes does not include a person, who, *without participating in the management of a facility, holds indicia of ownership primarily to protect a security interest in the facility.*

CERCLA does not define the phrases "indicia of ownership," "primarily to protect a security interest," or "participation in the management of a facility." It is reasonable to conclude, however, that a conventional oil and gas mortgage and loan agreement is no more than indicia of ownership primarily to protect a security interest, and that the conventional covenants incident to an oil and gas loan do not constitute a participation in management. This includes such covenants as the agreement to:

- Provide periodic reports of operating data
- Provide periodic reports of accounting data
- Conduct operations in compliance with laws and administrative regulations
- Maintain leases and contracts in good standing
- Maintain properties in good repair and condition
- Operate leases in a good and workmanlike manner
- Comply with lease covenants, including payment of royalties
- Pay charges for operations and protection against liens
- Pay taxes
- Maintain insurance
- Allow lender access to the property for inspection and testing
- Maintain accurate records and books

Insofar as the covenants involve environmental regulation, at least some care should be taken in exercising the lender's rights under the usual provision allowing the mortgagee, at the mortgagor's expense, to perform an act required by the mortgagor's covenants. The courts generally have recognized the right of the mortgagee, without loss of the secured creditor's exemption, to protect the collateral in any manner that is in accordance with accepted commercial practice. At least one court, however, has suggested that a lender may become liable for a borrower's environmental violations if he undertakes control of the borrower's environmental compliance *or has the power* to exercise such control.[3]

Although a present assignment to the lender of proceeds from the sale of production may be characterized as more than a lien interest, as such, even under the strictest decisions of the conflicting judicial interpretations of the secured creditor exemption in CERCLA, there is no indication than an assignment of revenues is such a participation in the management of a facility as to make the exemption inapplicable.

A covenant in many oil and gas mortgages allowing the mortgagee to go into possession in event of default or threat of default is a different question, however. The mere existence of such a covenant in a mortgage may not of itself constitute a participation in management. However, if the lender takes possession of the collateral by virtue of the covenant, this action almost certainly will be considered outside of the secured creditor exemption and will subject the lender to CERCLA liability as an "operator."

The extent to which the lender can police the loan when he feels repayment is impaired, short of actually going into possession of the collateral, is uncertain under the judicial decisions. The only generality that can be offered is that the courts tend to decide the issue on a case-by-case basis, depending on the facts of the particular case.[4]

The lender's exposure under CERCLA becomes critical in the event of foreclosure. Legally, if the lender is the buyer at the foreclosure sale, he becomes an owner. As a practical consequence, there has been a historical reluctance of lenders to proceed with foreclosure simply because of CERCLA liability exposure.

On April 29, 1992, after extensive notice and comment, the EPA (Environmental Protection Agency) promulgated a formal administrative rule concerning the secured lender's CERCLA liability. In very general summary, significant features of the EPA's rule are that:

- "Security interest" and "indicia of ownership" cover all legitimate financing transactions, without regard to type, form, or nomenclature.
- "Participation in management" means actual control (as distinguished from merely influencing control or possessing an unexercised power to control) either (A) of borrower's environmental compliance or (B) of all of borrower's operational affairs (as distinguished from financial or administrative affairs).
- Foreclosure does not automatically result in loss of the exemption if the lender acts to sell the property promptly,

but there is no limit on the length of time that the property can be held if no offers of fair consideration are tendered.

- The lender's continuing operation after foreclosure in a manner consistent with accepted commercial practice does not result in liability under CERCLA as an operator.

The EPA's rule was widely hailed as a resolution of many of the conflicts in the law and as affording relief generally favorable to lenders. On February 4, 1994, however, in what is commonly referred to as the *Kelley Decision,* the Circuit Court of Appeals for the District of Columbia Circuit held the EPA's rule invalid on procedural grounds.[5] Under the circumstances, a more lengthy discussion of the detailed provisions of the EPA's rule becomes little more than an academic exercise.

As stated by one banking organization, the *Kelley Decision* spread confusion once again throughout the banking industry. The lender is left to seek protection against the risk of environmental liability only by conventional means such as investigation of the collateral, environmental reviews and audits, covenants of the borrower, representations and warranties of the borrower, and insurance.

Louisiana

Under Louisiana's "mini-CERCLA" statute, a lender is protected from liability even if he becomes an owner or operator of property by virtue of foreclosure or *dation en paiement* (a voluntary "giving in payment" by the debtor) unless the lender causes the discharge or disposal of a hazardous substance, or knows at the time the security interest becomes enforceable that the property contains a hazardous substance.[6]

Mid-Continent

It should be noted that a lender can also face exposure under state statutes, such as the Kansas Hazardous Waste Act, although no reported decisions yet exist holding lenders liable under the act. Additionally, the Kansas Storage Tank Act provides that a lender may be liable under its provisions if he takes possession of a petroleum tank through foreclosure or other related activities.

Rocky Mountain

Utah's Hazardous Substances Mitigation Act concerning cleanup of hazardous waste sites exempts from liability for cleanup of contaminated properties lenders and persons acting as fiduciaries or custodians who have indicia of title but do not participate in the management of a facility.[7] This exemption is thus broader than the EPA's Lender Liability Rule. Utah also has amended its "One Action Rule," which normally requires a lender to foreclose first on real property collateral before proceeding against the borrower, to allow a lender to waive his rights to foreclose against non-residential real property and proceed against the borrower if the property is environmentally impaired.[8]

CHAPTER 7 ENDNOTES

1. See the discussion of CERCLA liability in the previous chapters.
2. CERCLA § 101(20)A; 42 USC § 9601(20)A. The full text of the applicable portion of CERCLA § 101(20)A is contained in Appendix 2–7.
3. See case and comment cited *infra* note 4.
4. Refusal of the United States Supreme Court to act on writ of certiorari to the decision in the leading case of United States v. Fleet Factors Corp., 901 F.2d 1550 (11th Cir. 1990), leaves the ultimate issue unresolved. There is disturbing language in the Eleventh Circuit opinion in Fleet Factors that the right of the lender, if it chooses, to affect the borrower's waste disposal decisions is a sufficient involvement in management control to create CERCLA liability for the lender.
5. Kelly v. EPA, 25 F.3d 1008 (D.C. Cir. 1994).
6. La. R.S. 30:2277(4).
7. Utah Code Am. § 19–6–302(19)(c).
8. Utah Code Ann. § 78–37–1.5.

8

Offshore Transactions

The offshore waters of the United States offer a unique and intriguing world for oil and gas exploration and production. The technical and operational challenges are obvious. Equally challenging, however, are the legal and regulatory aspects of conducting operations in this area, particularly those pertaining to environmental regulation.

Since a great deal of environmental regulation applies with equal force to offshore and onshore transactions, other chapters of this book also must be considered for offshore transactions and operations. This chapter is basically limited to special provisions and applications of the general statutory and regulatory schemes where they are unique to offshore transactions and operations. Special state environmental regulations that apply to coastal waters, such as regulation stemming from state coastal zone management acts, are beyond the scope of this chapter.

Investigation by Potential Buyer

The buyer's investigation of environmental liability exposures prior to an acquisition (sometimes loosely referred to as a due diligence review) is discussed in detail in Chapter 21. For offshore acquisitions, the investigation is modified only in its emphases. For instance, as a result of their location, offshore facilities do not have pits or similar storage or disposal areas. On the other hand, special environmental programs closely regulate disposal into the water of wastes and contaminants generated by operations.

Furthermore, the occasions for release of hazardous substances into the environment, as proscribed by CERCLA (Comprehensive Environmental Response, Compensation, and Liability Act) may be less prevalent in offshore operations, and therefore CERCLA liability may be less of a threat. Nevertheless, a thorough review of the property and the historical operations that have taken place on it remain no less critical.

Particularly, the permit file review should address the enforcement history for the facility, including any consent or compliance orders for improved performance that would transfer with an assigned permit or authorization. For example, most Gulf of Mexico facilities operate under a general NPDES (National Pollutant Discharge Elimination System) permit for discharges, but the EPA (Environmental Protection Agency) is authorized to require individual permits in certain circumstances. The individual permit may have different discharge parameters (possibly more stringent) than the general permit.

While a conventional title search is unnecessary for federal outer continent shelf (OCS) leases, nevertheless the buyer should ascertain the identity of all previous lessees. Since operating rights in the leases quite commonly are transferred without formal assignment of the leasehold estate, the buyer should ascertain the identity of all previous operators.

Any site inspection performed by an independent consultant should include an underwater survey and assessment. This assessment serves several important purposes:

1. It provides an environmental baseline of the property at the time of sale for both the seller and the buyer, and with regard to third parties.
2. The assessment may confirm the accuracy of the seller's warranties and representations.
3. It may indicate the condition of pollution treatment or operating systems—for instance, how recently tanks have been cleaned, whether the tank bottoms were handled as non-exempt hazardous waste or exempt waste under the RCRA (Resource Conservation and Recovery Act), and whether there were any sample analyses supporting the waste characterization and designation.
4. It may reveal whether any corrective action may be necessary due to past operations, as, for instance, contaminated sediment beneath the platform.

The site assessment also may indicate whether the chemical and hazardous substance inventory on the facility is supported by appropriate documentation and whether appropriate emergency prevention and response plans have been prepared as required by the Clean Water Act (CWA), the Oil Pollution Act of 1990 (OPA) and regulations under these acts, as well as special regulations issued by the Minerals Management Service (MMS).

If a property is located in offshore waters where exposure to oil spill liability for natural resources damages is deemed to be significant (for instance, in ecologically sensitive areas), the site assessment should include information that creates a baseline of data for a natural resources damages assessment, should it later become necessary.

Statutory Environmental Liabilities

Almost every federal environmental statute and the regulations under it affect both offshore and onshore operations. There may be special provisions in any of these acts, however, which apply only to offshore operations. Although the purpose of this chapter is to highlight such special provisions, at the risk of repeating discussion in other chapters, the general purposes of the various statutes are mentioned in order to put into context the special offshore applications.

Clean Water Act. For offshore transactions, the primary federal environmental statute affecting operations is the Clean Water Act (CWA). The CWA prohibits the discharge of pollutants into waters within the jurisdiction of the United States without a permit.[1] The statute regulates the handling, treatment, and discharge of numerous wastewater and other waste sources from offshore exploration and production activities. These include drill fluids and solids, produced water, sanitary and domestic waste, deck drainage, well treatment, completion and workover fluids, and produced sand through effluent guidelines defined for the offshore oil and gas extraction industry.[2] These effluent guidelines form the basis for discharge parameters contained in the NPDES permits for these discharges. These permits may be for an individual facility or as in the Gulf of Mexico for a general permit applicable to similar facilities discharging similar pollutants in a specified geographic region.

The operator's coverage under a NPDES general permit can be obtained by submitting a written notification to the EPA Regional Administrator within fourteen days prior to the commencement of the discharge that is to be covered. Unless otherwise notified by the Regional Administrator in writing, the operator is authorized to discharge in accordance with the general permit.[3] If the operator ceases to use the general permit, he must notify the EPA Regional Administrator within 60 days after the operator's permanent termination of discharge from its facility.[4] Coverage under individual permits can be transferred from seller to buyer but requires written notification to the EPA of the intended transfer at least 30 days prior to it. Close coordination of the transfer of permit status should be included in closing activities of a sale, so that there is no lapse in permit coverage and the proper permittee is responsible for discharges.

Prior to an acquisition, investigation should be made as to whether the facility being acquired has had difficulty in meeting its permit limits with existing equipment by checking both facility compliance records and agency enforcement records. The investigation should determine whether there are unpermitted discharge streams. The condition of the receiving waterbottom in the vicinity of the discharge point should be investigated to determine whether there is any accumulation of sludge or other material that may require removal.

It is also important to determine if the facility is located in or near a MMS defined "no activity" area (which may affect discharge rate) or included within an EPA "no discharge" prohibition under its permit with respect to a particular waste stream (e.g., oil-based drilling fluids). In a "no discharge" zone, the wastes must be barged to shore for treatment and disposal, and may be subject to potential RCRA or CERCLA liability for improper handling or disposal. Alternatively, potential equipment or technology upgrades (or both) may be necessary for efficient management of the waste on-site, as, for example, disposal into an injection well. While stormwater discharges are a new focus for onshore facilities, traditionally such discharges have been regulated by the "deck drainage" provisions of offshore NPDES permits and should not require an additional permit.

Clean Air Act. The CAA, both prior to its major amendment in 1990 and subsequent to it, applies to both offshore and onshore operations.[5] General application of the Clean Air Act to oil and gas exploration and production is discussed in detail in Chapter 17. CAA regulatory authority for federal OCS leases, however, is divided between the MMS and the EPA.

Since the 1990 CAA amendment, the MMS (which originally had sole CAA regulatory authority over all federal OCS leases), has had such authority over only federal OCS leases in the OCS area located in the Gulf of Mexico offshore from Texas, Louisiana, Mississippi, and Alabama. Even in these areas, however, the MMS is under statutory mandate to consult with the EPA to determine whether additional controls are necessary.[6]

In regard to all other offshore areas, CAA regulatory authority is vested in the EPA and is subject to consultation with the MMS.[7] For sources located within 25 miles from the coastward line of the OCS, however, the regulatory standards are identical with those specifically approved by the EPA for the adjoining state. See 40 CFR, Part 55.

RCRA. The RCRA regulates the universe of solid waste and its subset, hazardous waste. The statute directs the EPA to categorize as hazardous any waste that can be a significant hazard to human health and the environment if improperly handled. This is done by the EPA, which lists the waste per se as hazardous, or by the generator, who evaluates the waste according to specific hazard characteristics (e.g., ignitability, corrositivity, reactivity, and toxicity). If a waste is classified by RCRA as hazardous, it is subject to stringent regulation and requirements for its handling, storage, transportation, and disposal.

Historically, the RCRA has had a lesser impact on exploration and production activities because of an exemption that is commonly referred to as the RCRA oil field exemption, which is more fully discussed in Chapter 11. It is important to emphasize that not all wastes generated in exploration or production operations are covered by the exemption. Examples of non-exempt wastes include waste lubricants, hydraulic fluids, paint, waste solvents from cleanup operations or equipment maintenance, off-spec and unused material intended for disposal, filters, waste iron sponge, glycol and other separation media, and drums whose contents are not intended for use.

Non-exempt hazardous waste generated in offshore operations requires onshore disposal, as may certain exempt hazardous waste and nonhazardous waste, including produced water and most drilling fluids and cuttings. Offshore operations can create greater exposure to RCRA liability due to the temporary storage of hazardous waste at the platform longer than the period allowed by regulations and due to the increased hazard of transportation to an onshore treatment, storage, and disposal facility.

CERCLA. CERCLA regulates both the release of hazardous substances into the environment and the consequent necessary cleanup to protect human health and the environment. The definition of a CERCLA hazardous substance is very broad, but specifically *excludes* crude oil and any fraction thereof and natural gas, natural gas liquids, liquefied natural gas, and synthetic gas usable for fuel.[8] Releases of crude oil or natural gas or the resulting contamination, therefore, are not subject to regulation under CERCLA.

Nevertheless, normal lease operation usually involves use or generation of hazardous substances that are not crude or natural gas. Therefore, there is potential CERCLA liability exposure in lease operations and particularly, in the case of offshore operations, hazardous substances that may have been deposited in bottom sediments. Thus, a thorough bottom survey is necessary. Although certain wastes from exploration and producing operations are exempt under the RCRA definition of hazardous wastes, as discussed above, if the waste is neither crude oil nor natural gas and if it is specifically designated as a CERCLA hazardous substance, its release may give rise to CERCLA liability. It is important to note that the CERCLA petroleum exclusion and the RCRA oil field exemption do not coincide perfectly.

As with the RCRA, there is a similar potential for liability under CERCLA for disposal of CERCLA hazardous substances at onshore disposal sites.

Liability under CERCLA for removal and remediation costs, including natural resource damages, caused by the release of a CERCLA hazardous substance: (1) is strict (i.e., it arises even in the absence of negligence or intentional wrongdoing); (2) is retroactive to acts occurring prior to the passage of CERCLA in 1980; and (3) is joint and several among two or more defendants, i.e., any one of them is liable for the entirety of the cleanup costs and not simply for a part based on the individual defendant's proportionate contribution to the pollution. CERCLA liability exists for the current owner or operator of the contaminated site and any prior owner or operator of the property at the time the hazardous substance was released. Even a prior interim owner or operator who acquired the property from an owner or operator who did not cause the previous release of the hazardous substance may be liable as a result of failing to disclose actual knowledge of the release or condition. Further, liability attaches to anyone who arranged for the disposal or treatment or who transported a CERCLA hazardous substance.

CERCLA also imposes an obligation for release reporting of unpermitted releases exceeding a designated threshold. Thus, an investigation of operating records must be made to determine whether such reporting has been performed or was necessary.

Oil Pollution Act. Prior to 1990, the liability for the costs of cleanup of a release into offshore waters of oil or petroleum products exempt from CERCLA was principally governed by the CWA and applicable state law. The enactment in August 1990 of the Oil Pollution Act (OPA) imposed new strict and joint and several liability for owners and operators of offshore facilities (including the lessee or permittee of the area in which the facility is located) for damages to third parties and for the costs of removal of an unpermitted discharge of oil, as defined, into the navigable waters, including the territorial sea.[9] Therefore, if offshore facilities are being purchased, the due diligence site inspection or assessment previously mentioned should also include inquiry into past and current releases, or threatened releases of oil from such facilities.

In addition to imposing strict liability on owners and operators of offshore facilities for the damages and costs of removal of oil spills, the OPA established financial responsibility requirements that owners or operators must provide to insure payment for the OPA liabilities (up to $150 million for offshore facilities).[10] Although the seller's compliance with the financial responsibility requirement is not particularly the buyer's concern, the buyer must determine his own financial responsibility requirement and the approved manner in which it will be accomplished. This can result in a significant added cost or financial restraint on the purchase and may materially affect the future economic viability of the facility.

Of immediate concern to the buyer of an offshore lease are amendments to CWA enacted as part of the OPA. These amendments add a requirement that the owners or operators of vessels or offshore facilities may not handle, store, or transport oil unless: (1) a required response plan has been prepared, submitted to, and approved by the appropriate federal authority, that is, the U.S. Coast Guard, the EPA, or the MMS, as the case may be, and (2) the vessel or facility is operating in compliance with the plan.[11] To enforce these requirements, CWA establishes a broad range of administrative and civil penalties that can be applied for violations of the response planning requirements and for discharges, in addition to substantially increased criminal penalties. The law also grants federal agencies broad inspection powers, including the obligation to hold without prior notice inspections and periodic drills of oil removal capabilities.

As a result of these new laws and regulations, the buyer of an offshore facility must determine what spill prevention and response plans have been prepared by the operator, whether the plans have been submitted and approved as required, and whether the facility is operating in compliance with the plan. The seller should be required to provide the buyer with all relevant information relating to the plans, approvals, drills, inspections, and penalties, including any conditions and agreements requested by the governing agency. As part of the due diligence review, the buyer should also go to the governing agency and ask to review the facility's files, to confirm the facility's right to continue to handle, store, or transport oil and the prior history of its violations.

The buyer must become familiar with the specific requirements of an appropriate response plan for that type of vessel or facility and the corresponding operational and financial impact. In general, the plan should include the following:

- Full details of the response methods.
- How the plan relates to other approved contingency plans.
- Procedures for early detection of oil spills.
- Timely notification, numbers, types, capacities, and availability of the oil spill contamination equipment necessary to implement the plan.

It should also include the following:

- Number, training preparedness, and fitness of all full-time prepositioned personnel assigned to direct and implement the plan.
- Periodic training and drill programs to assure that personnel and equipment are in a ready state.
- The means of identifying, protecting, and mitigating effects of the environment and natural resources.

- Arrangements for the prepositioning of oil spill equipment and response personnel.
- The storage, transportation, and disposal of recovered oil.
- Personnel and third party safety.

The overall objective of response plans should be to remove as promptly and properly as possible the spilled oil and to minimize environmental damage from a worst-case discharge.

Naturally Occurring Radioactive Material (NORM)

A relatively recent discovery in oil and gas operations is the presence of NORM in active oilfield equipment and waste streams. Uranium, radium, thorium, and other radioactive elements and their decay products can be contained in some geologic formations from which oil, gas, and other minerals are produced. When hydrocarbons and formation water are produced, traces of these elements can be carried along and deposited in scale or solid residue in equipment such as tubing, pipe, filters, and separators, or they can be present in produced water streams. It is important to keep in mind, however, that NORM does not generally occur in all production, but only in production from particular and identifiable geographic areas.

Currently, NORM is regulated primarily by the individual states. A more detailed discussion of NORM is found in Chapter 12. Thus, a review of applicable state regulatory provisions must be made before removal of NORM or NORM-contaminated equipment from OCS offshore facilities to a particular onshore location.[12]

Discharge of traditional oilfield waters can be affected by the presence of NORM, although to a lesser degree than in equipment. The EPA has reviewed existing operating data from Gulf of Mexico facilities with respect to the presence of NORM in produced water for the purpose of establishing effluent guidelines for offshore NPDES permits. It has determined, however, that it does not have sufficient data concerning the presence of radionuclides across the entire offshore subcategory to regulate NORM at this time.[13] The EPA is continuing to gather information to re-evaluate this position.[14]

The offshore effluent guidelines also prohibit the discharge of produced sand. While this action was not primarily due to the presence of NORM, the EPA was influenced by the MMS action in setting stringent parameters for produced sand discharges in the

Gulf of Mexico based on NORM concentrations.[15] Therefore, produced sand is required to be disposed of onshore. The degree to which such onshore disposal is regulated depends on the laws of the state selected for the disposal.

Lease Requirements

In a private lease relationship between nongovernmental parties, it is uncommon for the lessor to impose more stringent environmental conditions on the lessee than those imposed by general law. However, such is not the case for federal leases. The Outer Continental Shelf Lands Act gives the MMS wide latitude to protect the environment.[16] The MMS has developed regulations to implement that authority in 30 CFR Part 250, Subpart C.

Generally, the obligations of these regulations fall on the lessee, defined as the "party authorized by a lease, or an approved assignment thereof, . . ." However, an operator can be designated to act on behalf of the lessee. There are detailed requirements for approval of a lease assignment by the MMS, including consultation with the U.S. Attorney General. Of primary importance is the fact that the assignor remains fully liable for all lease obligations prior to the approval of the assignment. Since there is no time limit for the review of the assignment by the MMS, provision should be made in the sales agreement to address this problem.

In addition to the requirements of the EPA-implemented environmental programs, such as CWA, RCRA, and CERCLA, pollution occurring from operation of an OCS facility must also be controlled or prevented to the satisfaction of the MMS at the expense of the lessee, and proposed corrective action plans must be reviewed by the MMS.

The MMS has express authority to restrict or control discharges that can cause unreasonable degradation of the marine environment. A due diligence review should include examination of the MMS files for operating restrictions in the area, such as the Flower Garden Banks, or for specific discharge materials, such as NORM in produced sand.

Specifications for facility design must also incorporate environmental requirements. For example, the regulations impose specific requirements for curbs, gutters, drip pans, and drains to prevent the discharge of contaminants into offshore waters. A site inspection must be made to determine if the facility has met this obligation or has been cited by the MMS for failure to comply.

There is a general prohibition against disposal of equipment, cables, chains, containers, or other materials into offshore waters. Similarly, there are requirements for handling and labeling materials, equipment, rods, containers, or other items that are likely to snag or damage fishing devices.

Finally, the buyer of an offshore facility should be aware of specific requirements for well abandonment procedures, including clearing the location of wellheads, casings, pilings, and other obstructions to at least 15 feet below the mudline. A site clearance verification must be filed with the MMS.

Platform and well abandonment can also trigger compliance issues with the Endangered Species Act and Marine Mammal Protection Act. Civil and criminal liability can result under each statute for the unpermitted "taking" of a species. The Endangered Species Act is discussed in greater detail in Chapter 14. The MMS site clearance plans can require biological monitoring for the presence of these species and specific removal procedures to minimize potential impact on them.

MMS *Requirements*

Once the federal OCS lease is assigned, the MMS has the authority to impose more stringent environmental conditions on the assignee than are expressly required by the lease. The buyer should discuss with the MMS the likelihood of increased future obligations arising out of the assignee's operation of the lease and may negotiate concerning them.

California

In California, the regulation of oil and gas operations in state waters involves to some degree a different regulatory structure than that applicable to onshore operations. That regulatory structure is discussed in the California comments to Chapter 9 (plugging abandoned wells) and Chapter 10 (oil spill cleanup).

As this chapter correctly notes, air quality standards on the Outer Continental Shelf within 25 miles of the coast must be updated periodically to remain consistent with the requirement of the corresponding onshore area. Therefore, an OCS oil and gas operator must be concerned with California's onshore air regulations, which are discussed in the California comments to Chapter 17.

Louisiana

As previously mentioned, a variety of wastes generated offshore can be disposed of at onshore locations. For Louisiana waste site destinations, hazardous waste shipments require a manifest prepared in accordance with the Louisiana Department of Environmental Quality's Hazardous Waste Regulations.[17] In addition, nonhazardous or RCRA-exempt oilfield waste shipments require a manifest prepared in accordance with the requirements of Office of Conservation Order 29-B. While there have been differences of opinion in the past regarding whether the manifest should originate from the offshore point of generation or from the onshore receiving dock, the more conservative view is that it should originate at the offshore location.

CHAPTER 8 ENDNOTES

1. CWA § 301; 33 USC § 1311.
2. Oil and Gas Extraction Point Source Category; Offshore Subcategory Effluent Limitations Guidelines and New Source Performance Standards, 58 Fed. Reg. 12,454; 40 CFR Part 435 (1993).
3. Final Modification to NPDES General Permit for the Western Portion of the Outer Continental Shelf (OCS) of the Gulf of Mexico, 58 Fed. Reg. 63,964 & 63,966 Part I.A.3 (1993).
4. *Id.* at 63,966 Part I.A.3.
5. 43 USC § 1334(a)(8).
6. CAA § 328(a); 42 USC § 7627(b).
7. CAA § 328(b); 42 USC § 7627(a).
8. CERCLA § 101(14); 42 USC § 9601(14).
9. 33 USC § 2701–2761.
10. OPA § 1016(c); 33 USC § 2716.
11. CWA § 311(j)(5); 33 USC § 1321.
12. See for example, Louisiana regulations for NORM. La. Admin. Code tit. 33:XV, §§ 1401, et. seq.
13. Oil and Gas Extraction Point Source Category; Offshore Subcategory Effluent Limitations Guidelines and New Source Performance Standards, 58 Fed. Reg. 12,454 & 12,483 (1993) (to be codified at 40 CFR Part 435).
14. Final NPDES General Permit for the Western Gulf of Mexico Outer Continental Shelf, 57 Fed. Reg. 54,642 & 54,653 (1992).
15. Oil and Gas Extraction Point Source Category; Offshore Subcategory Effluent Limitations Guidelines and New Source Performance Standards, 58 Fed. Reg. 12,454 & 12,467 (1993) (to be codified at 40 CFR Part 435).
16. 43 USC §§ 1334, 1346, 1348.
17. LAC 33:V.901 et seq.

9

Plugging Abandoned Wells

Oil and gas operators are no strangers to environmental controls of plugging and abandoning operations at a state level. In a 1987 report to Congress, the EPA (Environmental Protection Agency) documented that, notwithstanding regulations, many states have tens of thousands of unplugged or improperly plugged abandoned wells, which have caused groundwater chloride contamination.[1] In addition to saltwater incursion into freshwater horizons, of course, the problem involves the migration of down-hole hydrocarbons up the wellbore to the wellhead and out onto the surface. There are, however, no federal rules or regulations specifically covering plugging and abandonment. This remains a problem solely for state control.

In Texas, the statutory requirements[2] are administered by the Railroad Commission of Texas (RRC) under Statewide Rule 14.[3] The responsibility for plugging the well belongs to the operator of the well or property as shown by the RRC records. Under present RRC practice, this is determined from the RRC Form P-4 (entitled "Producer's Transportation Authority and Certificate of Compliance"), that is currently on file. An operator is liable whether or not he owns a working interest in the well. If the RRC cannot find the operator to issue an order requiring the plugging or if an imminent danger of pollution exists, the RRC can plug the well and recover the cost of plugging plus penalties from the operator. Amounts due to the RRC are secured by a statutory lien on the property.

If the operator cannot be found or does not pay, the RRC can recover the cost of plugging from the owner or owners of a non-operating working interest at the time the plugging obligation accrued. If there is more that one non-operating working interest owner, their liability is not joint and several as among them; each is liable only for a part of the cost proportionate to his interest in the well or property.

It is the position of the RRC, supported by the holding in a Texas Court of Appeals case,[4] that a non-operating working interest owner is liable even though he opted out of participation in the well under applicable provisions of a joint operating agreement. Amendment of the statute in 1993, however, casts doubt on the validity of this position.

Neither the landowner nor the owner of a royalty interest or overriding royalty interest has any liability under the plugging statute.

In the event of sale of a well or property, the current statute provides that the selling operator can be absolved from further plugging liability if the well in question was in compliance with applicable RRC rules at the time of the sale (for instance, the well was not leaking) and if prescribed procedures are complied with, including the execution of a Form P-4 by the buyer as the new operator, and its approval by the RRC. The Form P-4 currently in use includes the following language in distinctive red type:

> *By signing this certificate as the CURRENT OPERATOR, I acknowledge responsibility for the regulatory compliance of the subject lease, INCLUDING PLUGGING OF WELLS if required under Statewide Rule 14. I also acknowledge that I will remain designated as the Current Operator until a new certificate designating a new Current Operator is approved by the Commission.*

As a consequence, the buyer of a property can anticipate that the seller reasonably may require as a condition of closing the execution and filing of an appropriate Form P-4 and its approval by the RRC.

Especially in recent years, the RRC has shown a readiness to refer cost recovery and penalty cases to the Texas Attorney General for prosecution through the courts. The Attorney General has taken a particularly aggressive position that officers and directors of a corporation are personally liable for plugging costs and penalties if the corporation is no longer in good corporate standing, for instance because of failure to pay franchise taxes.

Failure properly to plug abandoned wells can have consequences beyond payment of costs and penalties required by the RRC. Operators who are guilty of unresolved violations (including unpaid fines) are not allowed to receive new drilling permits.[5] Furthermore, an operator who has experienced environmental related problems (even if corrected) during the previous four-year period may be required to post a bond or furnish other evidence of financial responsibility rather than simply paying the $100 annual filing fee.

A practical problem exists concerning well-plugging obligations, particularly in older fields, in that the existence and location of abandoned wells may not presently be known, even to the RRC. Thus, prior to a purchase, investigation as to the existence of abandoned wells should be made from all possible sources, including inquiry on the ground.

California

In California, the requirements for plugging and abandoning wells are administered by the Division of Oil and Gas and Geothermal Resources of the Department of Conservation. Formerly, it was simply known as the Division of Oil and Gas and is still commonly referred to as the "DOG." The DOG is headed by the "Supervisor." The law requires the owner or operator to file a written notice of intention to abandon a well with the Supervisor of the DOG.[6] The notice of intention must contain the total depth of the well to be abandoned, the complete casing record of the well, and such other pertinent data as the Supervisor may require.[7] The regulations set forth the abandonment requirements in more detail. The Supervisor has issued detailed onshore abandonment requirements[8] and detailed offshore requirements.[9]

In addition, the Supervisor may order the abandonment of any well that has been deserted. Suspension of drilling operations and removal of drilling machinery is prima facia evidence of desertion after six months unless a request for an extension of time is granted.[10]

Whenever ownership or operation of any well is transferred from one party to another, the transferor must notify the Supervisor in writing in such form as the Supervisor may direct[11] and the transferee must also notify the Supervisor in writing, also in such form as the Supervisor may direct.[12]

There are also various bonding requirements,[13] and any owner or operator of a marine facility (which includes oil platforms, etc.) must apply for and obtain a certificate of financial responsibility.[14]

If the Supervisor deems a well to be hazardous, he may require the last operator who had an economic interest in the well to abate the problem. The Supervisor can also require a present surface owner or mineral estate owner to do the same if they derived a substantial financial gain from the well. If no such responsible person can be located, the Supervisor can classify the well as idle/deserted and perform the necessary work himself. There is a fund available to the Supervisor for such purposes, which is paid for by a modest tax on oil and gas producers in the state. However, the fund is quite small and the Supervisor may not make expenditures for this purpose that exceed the sum of $500,000 in any one year.[15]

In addition, there is a special provision giving the Supervisor power to order the reabandonment of any previously abandoned well when the future construction of any structure over or in the proximity of the well could result in a hazard. The cost of such reabandonment operations is declared to be the responsibility of the owner of the property upon which the structure will be located. However, if the well was not abandoned in accordance with the requirements of the DOG in effect at the time of the abandonment, then the last operator having an economic interest in or receiving any benefit from the well, if still in business in this state, is responsible.[16]

Louisiana

Louisiana Office of Conservation Statewide Order 29-B sets out specific procedures for plugging abandoned wells and requires that they be performed by the last operator of record.

Under the 1993 Oilfield Site Restoration Law,[17] the Louisiana Department of Natural Resources (DNR) administers a public fund designated for the cleanup of abandoned or "orphaned" oilfield waste sites. Site restoration specifically includes the plugging and abandonment of wells. The DNR has already listed several thousand well sites as "orphaned," and therefore subject to cleanup with public funds. Otherwise, the law differentiates between "unusable oilfield sites" and "orphaned sites." An unusable oilfield site is one for which no continued useful purpose for oil and gas activity can be found and for which no "responsible party" can be located. A "responsible party" includes both the last operator of record and working interest owners, but specifically excludes the landowner. When a public cleanup is conducted of an orphaned site, the law directs the DNR to sue any responsible party, including former operators of record and their working interest owners, using an "inverse chronological order."

An important point is that at the time any Louisiana oil and gas facility is transferred, the new law also encourages the creation of a site-specific trust account. These accounts, the terms of which can be largely bargained between the transacting parties, are specifically designed to set aside sufficient funds for future site restoration. If all conditions associated with the trust account are satisfied, the transferring party is absolved of all future liability for site cleanup.

Finally, similar to Texas law, operators who fall out of compliance with this new cleanup program face stiff penalties, including the statewide withdrawal of operating authority for all wells in Louisiana.

Mid-Continent

The Oklahoma Corporation Commission (OCC) and the Kansas Corporation Commission (KCC) are charged with the responsibility for adopting rules for the plugging and abandoning of wells. In Oklahoma, any working interest owner and operator are jointly liable for plugging.[18] KCC rules provide those “legally responsible” shall be liable for plugging costs.[19] If those responsible for plugging cannot be located or are financially unable to plug a well, the OCC or KCC can plug the well. In these instances, funds are available for plugging and statutory liens exist in favor of those who plug a well against the person responsible for the plugging.[20] The standards for plugging are established by the OCC and KCC.

By statute, in Kansas, a well that is abandoned and not properly plugged is “deemed likely to cause pollution of any useable water strata or supply.”[21]

Rocky Mountain

The Rocky Mountain states all have regulations that establish plugging procedures for abandoned wells.[22] To date, the Rocky Mountain states have not focused as much as Texas on liability of non-operators for plugging costs, but provisions establishing trust funds for plugging of abandoned wells where responsible parties cannot be located are quite common,[23] as are provisions allowing the sale of abandoned equipment for salvage to help pay plugging costs.[24]

The Rocky Mountain states also have bonding or other financial assurance requirements to secure these plugging obligations.

Of particular note to purchasers of producing properties is a very aggressive attitude by the Wyoming Oil and Gas Conservation Commission to insure that bonding levels for dormant wells are set at appropriate levels when a new operator takes over operations. The purchaser must notify the Supervisor of the Commission of the transfer at least 30 days *prior to closing* and the Supervisor may require additional bonding. In order to approve the new operator, the Supervisor also may require the new operator to post an additional bond of $2.00 per foot for total footage of dormant wells exceeding 12,500 feet, or the operator must file a plan to plug all dormant wells within two years.[25]

CHAPTER 9 ENDNOTES

1. Regulatory Determination for Oil and Gas and Geothermal Exploration, Development and Production Wastes, 53 Fed. Reg. 25446 (1988).
2. The statutory requirements are found principally in Chapter 89 of the Texas Natural Resources Code.
3. 16 TAC § 3.14.
4. Railroad Commission v. Olin Corp., 690 S.W.2d 628 (1985).
5. 16 TAC § 3.5.
6. Public Resources Code § 3229.
7. Public Resources Code § 3230.
8. 14 California Code of Regulations §§ 1723, et seq.
9. 14 California Code of Regulations §§ 1745, et seq.
10. Public Resources Code § 3237.
11. Public Resources Code § 3201.
12. Public Resources Code § 3202.
13. Public Resources Code §§ 3204, 3205, 3205.1, 3205.2, 3205.5, 3206 and 14 California Code of Regulations § 1722.1.
14. 14 California Code of Regulations § 791.
15. Public Resources Code §§ 3250, et seq.
16. Public Resources Code § 3208.1.
17. La. R.S. 30:80 et. seq.
18. (O.A.C. 165:10–11–3).
19. K.S.A. ff 15–179.
20. 520.S. 310, et seq. and K.S.A. ff 55–180.
21. K.S.A. ff 55–179.
22. The pertinent regulatory provisions for the Rocky Mountain states are as follows:
    Colorado: Colorado Oil and Gas Conservation Commission Rules, Rule 317, 2 Colo. Code Reg. 404–1.
    Nebraska: Nebraska Oil and Gas Conservation Commission Rules, T.267, Rule 3–028.
    New Mexico: New Mexico Oil Conservation Divisions Rules, Rules 201–204.
    North Dakota: North Dakota Admin. Code 43–02–03–33 to 38.1.

Utah: Utah Admin. Code, Rule 649–3–24.
Wyoming: Wyoming Oil and Gas Conservation Commission Rules, Rule 318.

23. For example, Colorado has established an Environmental Response Fund for this purpose, Colo. Rev. Stat., § 34–60–124, and North Dakota has an Abandoned Oil and Gas Well Plugging and Site Reclamation Fund, N.D. Cent. Code § 38–08–04.5.
24. See N.D. Cent. Code §§ 38–08–04.8 and 04.9.
25. Wyoming Oil and Gas Conservation Commission Rules, Rule 304.

10

Oil Spill Cleanup

Soil contamination by hazardous substances, as discussed in Chapter 1, is covered by CERCLA (Comprehensive Environmental Responsibility Compensation and Liability Act). Since by statutory definition, however, crude oil is not a hazardous substance, soil contamination by it is not covered by CERCLA—unless it results from contamination by a mixture of a hazard substance and crude oil—however disproportionately small the amount of hazardous substance in the mixture may be. Whether on the basis of mere sight inspection or actual soil testing, it is common experience that a great deal of soil contamination is the result of simple crude oil spills, such as storage tank over-fills or leakage at lines or valves.[1]

Texas RRC *Rule* 8

In Texas, the Railroad Commission of Texas (RRC) Statewide Rule 8 for many years has mandated the general cleanup of crude oil spills. Historically, however, the RRC has seldom enforced Rule 8 to require crude oil cleanup. Under recent state legislation, the RRC has now promulgated new Rule 91, dealing with oil spill cleanup standards and procedures, effective January 1, 1994. The rule explicitly does not apply to spills prior to the effective date.

Scope of Texas RRC *Rule* 91

Rule 91 applies to the cleanup of soil contaminated by spills from exploration and production operations, including pipelines. The two key limitations on the scope of the rule are: it applies only to spills of crude oil and Rule 91 applies only to spills into soils in non-sensitive areas. It does not apply to spills in sensitive areas, such as surface water bodies, areas with shallow groundwater, or parks, wildlife refuges, or residential areas. Spills in these sensitive areas may require more extensive cleanup. The RRC determines cleanup requirements for spills in sensitive areas on a case-by-case basis.

Standards and Procedures

Rule 91 outlines the following steps for the cleanup of a crude oil spill into soil:

- Remove all free oil immediately.
- Delineate the affected area, both horizontally and vertically.
- Bring all soil containing over one percent by weight (10,000 parts per million) (ppm) total petroleum hydrocarbons (TPH) to the surface for remediation or disposal.
- Handle all soil containing over five percent by weight (50,000 ppm) TPH using special procedures to prevent storm water contamination.
- Achieve a final cleanup level of one percent by weight TPH as soon as technically feasible, but not later than one year after the spill. The operator may select any technically sound cleanup method that will achieve the final cleanup level.

Alternative standards and procedures may be approved upon written request. The responsible operator must demonstrate that the proposed alternative will provide equal or greater protection of the environment.

Rule 91 also outlines reporting requirements for cleanup activities. These build upon the reporting requirements under Texas RRC Statewide Rule 20 and vary depending on the size of the spill. For spills over 25 barrels, the operator must submit to the RRC analyses of samples representative of the spill site to verify that the final cleanup level has been achieved.

Relation to RCRA

Although most crude oil spills are exempt from regulation under the RCRA hazardous waste management programs, spills from crude oil transmission pipelines are not. The EPA's hazardous waste regulations may apply to the cleanup of spills from these pipelines if the contaminated soils fail the RCRA toxicity characteristic test for benzene or other toxic constituents.

Water Spills

It has been observed that produced water spills can be more widespread than crude oil spills, and they may be more environmentally devastating simply because over time, spilled hydrocarbons will degrade naturally but salt plumes from produced water spills can endure indefinitely. Since at this time a RRC statewide rule has only been proposed, this rule is not discussed here in detail.

California

A number of California agencies have regulations that apply to oil spill cleanups. To begin with, the authority to regulate all oil and gas operations in California is vested primarily in the Division of Oil, Gas and Geothermal Resources of the Department of Conservation, commonly known as the "DOG." The DOG enforces the "Oil and Gas Conservation Law."[2] This law applies to both onshore and offshore operations. The latter include the tide and submerged lands of California out to the Outer Continental Shelf where federal jurisdiction begins—generally three miles from shore. For tide and submerged lands the California State Lands Commission plays a regulatory role.[3] The DOG jurisdiction is derived from the state's police power to regulate oil and gas production. The State Lands Commission's powers are derived from the power of an owner to regulate operations on its land.

The Division of Oil and Gas requires that all "operations," including oil spill cleanup operations, be in accordance with good oilfield practices.[4] It also issues permits to drill and conduct various operations,[5] has the power to regulate in a number of areas specifically related to environmental protection,[6] regulates plugging and abandonment practices,[7] and as already described in the California comments to Chapter 9, issues and enforces regulations to eliminate hazardous or idle deserted wells.[8] Furthermore, it has the power to order reabandonment of any previously abandoned well when the future construction of any structure over or in the proximity of the well could result in the hazard.[9]

Other agencies are also involved in the oil spill cleanup process. For instance, disposal of oilfield waste, such as debris from cleanup of an oil spill, must conform with regulations issued by the State Water Resources Control Board and the various California Regional Water Quality Control Boards.[10]

California has also adopted a law that is analogous to federal RCRA and that, among other things, applies to disposal of debris from oil spill cleanup. It is known as the California Hazardous Waste Control Act (HWCA).[11] The HWCA applies to "hazardous waste" as that term is defined. The California definition of hazardous waste is broader than that of RCRA, and includes the disposal of petroleum-contaminated soil.

In 1984 the Toxic Pits Clean Up Act (TPCA), which is administered by the State Water Resources Control Board and the local Regional Water Control Board, was added to the HWCA. It is designed to ensure the closure of existing "surface impoundments" containing liquid hazardous wastes that threaten a potential source of drinking water.[12]

Another law commonly affecting oil spill cleanups is the Porter-Cologne Water Quality Control Act. It obligates dischargers to clean up or abate the affect of any waste discharge that poses a threat to groundwater or to other waters of the state.[13] Note that there is no requirement that the substance involved be "hazardous." Under the authority of this act, the various regional Water Quality Control Boards have published guidance policy documents relating to the disposal and treatment of hydrocarbon-contaminated soils.

California has also enacted laws dealing with oil and hazardous spill prevention and response. In September 1990, the Legislature enacted the Lempert-Keene-Seastrand Oil Spill Prevention and Response Act, which became effective immediately.[14] The act created a State Oil Spill Response Administrator and enacted various code provisions dealing with marine oil spill prevention, contingency planning, response, and liability.

From the oil operator's standpoint, one important feature of the law is that it requires any person who "causes or permits" oil to be discharged into state marine waters to immediately clean up and remove the discharge in accordance with all applicable contingency plans and orders. The requirement applies without fault. In addition, "responsible parties" (which includes the owner of the oil), are strictly liable for unlimited clean up and removal costs. Thus, the owners of the spilled oil, along with the owners and operators of the facilities and vessels that are the source of the spill, are responsible for clean up whether or not they caused or permitted the spill.[15] In addition, parties responsible for such a discharge must promptly report to the California Office of Emergency Services.[16] Any such spill also should be reported to the appropriate city or county hazardous materials administration agency, which is generally the local fire department.[17]

In addition, the act gives the oil spill response administrator and the State Lands Commission joint regulatory authority over various marine facilities including pipelines and offshore platforms located on lands leased from the State of California.[18] The term "marine facility" includes facilities located in the marine waters or in a location where a discharge could enter marine waters. Most oil-related facilities so located are deemed "marine facilities," including onshore marine terminals, offshore platforms, drill ships, semi-submersible rigs, and jack-up drilling rigs.[19]

A plethora of other California statutes also impose liability for oil spills. They include statutes that:

- impose unlimited and strict liability for discharges of petroleum and hazardous substances from vessels;[20]
- impose unlimited and absolute liability for any discharge or leak of oil, gas, or drilling waste into state marine waters;[21]
- make it a misdemeanor to discharge oil from a vessel into the navigable waters of the state;[22]
- impose a civil penalty of up to $6,000 per discharge and unlimited liability for cleanup costs for anyone who negligently or intentionally permits the discharge of any oil into any state waters;[23]
- impose liability for the unlawful or negligent destruction of any animal protected under state law (including destruction caused by discharge of pollutants into state waters);[24]
- prohibit the discharge of any petroleum product or refuse into state waters that is harmful to animal or plant life;[25]
- in the event of a discharge of petroleum, authorizes the State Fish & Game Department to clean up whenever the petroleum is likely to enter state waters and to recover all costs from the responsible person;[26]
- require anyone responsible for polluting state waters in a way that injures plant or animal life to remove any such pollution;[27]
- impose liability on anyone who discharges any substance harmful to animal or plant life into any state waters;[28]
- among things, prohibit the discharge of oil and other substances into the ocean or state waters;[29] and
- impose civil and criminal penalties to be levied upon any person who discharges petroleum or hazardous substance into any state waters.[30]

Oil and gas operators in California should also be aware that various counties and cities in the state where oil and gas are produced have their own set of regulations that may pertain to oil spill cleanup.

Exposure to environmental liability can also arise under California common law theories. The law of "nuisance" is an example. One who creates on his property that which the law defines as a nuisance or who acquires a property upon which a nuisance already exists is responsible for its abatement. It is no defense that the nuisance was created by a prior landowner.[31] The California Civil Code[32] defines a nuisance as:

> *"Anything which is injurious to health, or is indecent or offensive to the senses, or an obstruction to the free use of property, so as to interfere with the comfortable enjoyment of life or property, or unlawfully obstructs the free passage or use, in the customary manner, of any navigable lake, or river, bay, stream, canal, or basin, or any public park, square, street, or highway, is a nuisance."*

Most environmental harms would probably be considered nuisances. For instance, both air and water pollution have been declared by statute to be nuisances in California.[33] In addition, there is considerable case law holding that various types of pollution and contamination constitute nuisances. On the other hand, to be actionable, a nuisance must be the result of intentional, reckless, negligent, or extra hazardous activity.[34]

Another theory of common law liability for environmental harm such as oil spills is the theory of trespass. The concept of trespass is very similar to that of nuisance; however, there must have been some sort of physical damage to the property in question. Also, case law holds that the trespass must be intentional, reckless, negligent, or the result of an extra hazardous activity.[35] Typically, a trespass involving an oil and gas lessee is defined as one in which the lessee has acted in such a way as to exceed his rights under the oil and gas lease and, in the process, has inflicted physical damage upon the property.

Another theory concerning environmental damage such as might result from an oil spill is the theory of waste.[36] Waste is defined as willful conduct injuring the real property of another. Thus, an oil and gas lessee who has an oil spill and refuses to clean it up may, by willful conduct, commit waste upon the remainder interest of the oil and gas lessor. In the discretion of the court, treble damages may be awarded if waste is proven.

There have also been cases in which persons claiming environmental damage sought damages for emotional distress by alleging that the defendant intentionally inflicted the emotional distress upon them.[37]

Another basis for environmental liability due to oil spills and the like in California is the familiar theory of negligence. For instance, a case recently held that negligence was available where a defendant illegally discharged hazardous substances into the ground knowing that these substances would pollute the soil and the groundwater, and then failed to disclose the existence of this contamination when the property was sold.[38]

All of the foregoing causes of action are tort claims. Therefore, punitive damages are also available if the wrongdoer is shown by "clear and convincing evidence" to be guilty of "oppression, fraud, or malice."[39]

In addition to the tort claims described above, contractual claims are available. Typically, an oil and gas operator holds as a lessee under an oil and gas lease. Having assumed the obligations of the lease, the oil and gas operator is typically liable for such cleanup obligations as are specified in the lease. Failure to meet those obligations will give rise to a breach of contract case.

Louisiana

In Louisiana, soil contamination by crude oil is regulated primarily by the Department of Natural Resources, Office of Conservation. If a particular spill impacts the state's surface or underground waters, however, the Office of Water Resources within the Louisiana Department of Environmental Quality (DEQ) will also become involved. In addition, crude oil spills into any media are subject to a DEQ release reporting scheme.[40] Spills exceeding 1 barrel in a 24-hour period require reporting.

Although no formal rules exist, soil cleanup standards for spills on oilfield sites generally follow pit closure requirements contained in Office of Conservation Statewide Order 29–B.[41]

Mid-Continent

The OCC has published a document entitled "Guidelines for Responding to and Remediating Salt Water and Oil Spills," which specifies recommended actions to be taken in response to a spill. These recommended actions are the following:

1. Initiating actions to prevent further discharge or release;
2. Utilizing an appropriate contaminant system; and
3. Removal of all free fluids as soon as practical.

The guidelines also recommend specific procedures to follow for various types of spills (e.g., oil spills to soil).

The Kansas Corporation Commission (KCC) is responsible for the cleanup of pollution resulting from oil and gas operations. The "KCC-KDHE Coordination Manual on Clean-up of Oil Field Related Contamination" contains limited cleanup guidelines.

Rocky Mountain

A few of the Rocky Mountain states now have regulations that establish specific oil spill cleanup standards and procedures. Colorado's Oil and Gas Conservation Commission recently adopted new Rule 907, Spills and Releases; Reporting and Remediation and Rule 908, Management of E&P Wastes, which generally deal with cleanup procedures. These rules are effective as of June 30, 1995. The Commission also plans to issue an Environmental Guidance Document on Spill and Release Remediation in the near future.

Nebraska's Oil and Gas Commission Rules 3–022.01 to .11 also provide specific procedures and cleanup levels that are applicable for spills that do not leave the lease nor reach surface or groundwater. Spills that do not qualify for this exemption are under the jurisdiction of the Department of Environmental Quality and applicable federal agencies, but there do not appear to be any specific DEQ cleanup standards or procedures.

The New Mexico Oil Conservation Division also has issued remediation guidelines for use on federal, state, and fee lands, but the OCD states that compliance with the guidelines does not insure compliance with all applicable regulations.[42]

CHAPTER 10 ENDNOTES

1. This chapter is an adaptation of pertinent portions of article by Lori Wrotenbery, Director of Environmental Services for the Oil and Gas Division of the Railroad Commission of Texas, entitled Environmental Regulation in the Oil Field: Update on Railroad Commission Programs,

State Bar of Texas Oil, Gas and Mineral Law Section Report, December, 1993.

2. Public Resources Code §§ 3000, et seq., 14 California Code of Regulations §§ 1710, et seq.
3. Public Resources Code §§ 6801, et seq.
4. 14 California Code of Regulations § 1722.
5. Public Resources Code § 3203.
6. See, for instance, 14 California Code of Regulations § 770(a), 14 California Code of Regulations § 776.
7. Public Resources Code § 3208; 14 California Code of Regulations §§ 1723–1723.8.
8. Public Resources Code §§ 3250, et seq.
9. Public Resources Code § 3208.1(a).
10. 14 California Code of Regulations § 1775.
11. Health & Safety Code §§ 25100, et seq.
12. Health & Safety Code §§ 25208.4, et seq.
13. Water Code §§ 13000, et seq.
14. 1990 Cal.Stat. Ch. 1248.
15. Government Code §§ 8670.25 and 8670.27(a).
16. Government Code § 8670.25.5(a).
17. Health & Safety Code § 25507.
18. Public Resources Code § 8755.
19. Government Code § 8670.3(f).
20. Harbors and Navigation Code § 293.
21. Harbors and Navigation Code § 294.
22. Harbors and Navigation Code § 133.
23. Harbors and Navigation Code § 151.
24. Fish and Game Code § 2014.
25. Fish and Game Code § 5650.
26. Fish and Game Code § 5655.
27. Fish and Game Code § 12015.
28. Fish and Game Code § 12016.
29. Public Resources Code § 6873.
30. Water Code § 13350.
31. Civil Code § 3483.
32. Civil Code § 3479.
33. Health and Safety Code §§ 41700 and 4461.
34. Tint v. Sanborn (1989) 211 Cal.App.3d 1225, 1228.
35. Wilson v. Interlake Steel Co. (1982) 32 Cal.3d 229, 233–35, 185 Cal.Rptr. 280.
36. Code of Civil Procedure § 732.
37. Potter v. Firestone Tire & Rubber Co. (1993) 6 Cal.4th 965, 25 Cal.Rptr.2d 550.
38. The Newhall Land & Farming Company v. Superior Court (1993) 19 Cal.App.4th 334, 348–351, 23 Cal.Rptr.2d 377.
39. Civil Code § 3294(a).
40. LAC 33:I.3901 et seq.
41. LAC 43: XIX.129.B.
42. N.M. Oil Conservation Division Environmental Regulations, Tab 7b, at page 2.

11

Hazardous Wastes

RCRA *and the* Oil *and* Gas Exemption

The Resource Conservation and Recovery Act (RCRA)[1] (the purpose of which is more easily understood by the name of its predecessor statute, the Solid Waste Disposal Act) authorizes the EPA (Environmental Protection Agency) to regulate the management of wastes resulting from industrial, commercial, mining, agricultural, and community activities. Subtitle C of the RCRA contains a comprehensive program for the regulation of hazardous wastes, while nonhazardous wastes are regulated by Subtitle D. Produced waters, drilling fluids, and other wastes associated with exploration and production are presently exempt from regulation as hazardous wastes under Subtitle C.[2]

The exempt oil and gas wastes fall into three major categories:

- Produced waters—waters produced with and then separated from oil and gas
- Drilling fluids—mixtures of water, clay, barite, and other additives used in drilling wells
- Associated wastes—other wastes uniquely associated with drilling and production operations, such as crude oil tank bottoms

Produced waters make up about 98% of all oil and gas wastes. In Texas, about 98% of these produced waters are injected in wells regulated under a federally approved underground injection control program administered by the Railroad Commission of Texas. Drilling fluids plus associated wastes make up about 2% of oil and gas wastes.

The exempt oil and gas wastes are unique. They are high in volume, but relatively low in toxicity. Also, oil and gas wastes are generated in diverse operations and environmental settings. Exempt oil and gas wastes are regulated under state and federal programs that have evolved over the years to address these special circumstances. They are not amenable to overly prescriptive, inflexible regulation.

Recognizing the unique characteristics of oil and gas wastes, Congress in 1980 specifically exempted most oil and gas wastes from regulation as hazardous wastes under Subtitle C pending further study by the EPA.

In its regulatory determination, the EPA concluded that the exempt oil and gas wastes should not be regulated as hazardous wastes because such regulation: (a) is unnecessary, (b) would have a severe adverse economic impact on the industry, (c) would not allow the flexibility necessary to address the wide range of complex problems and management practices in different parts of the nation, and (d) would disrupt and duplicate existing federal state regulatory programs.

The EPA found that exempt oil and gas wastes pose no significant threat to public health or the environment when managed in accordance with existing federal and state programs. It determined that these existing programs are generally adequate for exempt oil and gas wastes, and that any necessary improvements should begin with them.

Scope of the Exemption

The EPA also defined the scope of the exemption. It determined, based on a review of the RCRA, that the exemption covers only those wastes uniquely associated with primary field operations. These specific types of exempt wastes are listed in Appendix 7, and nonexempt wastes are listed in Appendix 8.

It is sometimes believed, erroneously, that exploration and production operations generally are exempted from RCRA regulation. This simply is not true. Although the buyer of oil and gas properties need not be concerned about RCRA compliance by the seller as to wastes not covered by the exemption, he must determine which wastes from the properties are nonexempt and then take into account the associated cost and means of compliance with the RCRA. The buyer, of course, must also determine that arrangements for disposal of produced waters to comply with state regulatory requirements.

California

As mentioned in the California Comments to Chapter 10, California has adopted its own version of federal RCRA, known as the California Hazardous Waste Control Act (HWCA).[3] Like the RCRA, the California HWCA makes it unlawful to "manage" hazardous waste except as provided in the HWCA and the regulations adopted under that law by the Department of Toxic Substances Control, which enforces the act.[4] The term "management" of hazardous waste means everything that is done in connection with the handling, storage, treatment, disposal, etc., once a hazardous substance becomes a waste.[5] The HWCA defines hazardous waste as all wastes deemed hazardous under RCRA and additional wastes deemed hazardous according to California standards, even though they are not considered hazardous wastes under the RCRA. Most importantly to upstream oil and gas operators, the HWCA, unlike the RCRA, has no oil field production waste exemption.

In addition to a federally defined RCRA waste, a waste is hazardous under the HWCA if it has characteristics that are hazardous to human health or the environment if it is improperly managed.[6] The Department of Toxic Substances Control has adopted highly complicated regulatory criteria for identifying those hazardous wastes.[7]

An analysis of those criteria is beyond the scope of this book. Suffice it to say that most, if not all, produced waters (i.e., waters produced with and then separated from oil and gas) are classified as nonhazardous and are disposed of in so-called Class II wells. Such injection wells are permitted and regulated by the California Division of Oil and Gas.[8] A small amount of produced water is apparently disposed of in Publicly-Owned Treatment Works (POTW).[9] Drilling muds are classified as hazardous wastes if they contain additives that are themselves considered to be hazardous. As a practical matter, most California drilling fluids are disposed of in Class I landfills,[10] although some drilling fluid, particularly in oilfield cleanups, is treated by bioremediation at land treatment units, which in turn are regulated under the HWCA.[11] It should also be remembered that even if a waste is not classified as being hazardous, that does not mean it is nonpolluting. It simply means that it has not been singled out for special regulation as a hazardous waste.

There is a potential tax trap created by the California hazardous waste laws. Various taxes and fees are levied upon hazardous waste disposal facilities and those who dispose of hazardous waste.[12] California also has what is referred to as the "mixture" rule. Under the rule, if a nonhazardous waste is mixed with a hazardous waste, then the entire waste is deemed hazardous.[13] Thus, if one inadvertently mixes even a tiny amount of hazardous waste with a very large amount of nonhazardous waste—even if the hazardous waste is indistinguishable—then the State of California will also contend that the entire waste stream is hazardous and that the hazardous waste disposal fees should be calculated on the entire waste stream. Where the waste stream is large, as is generally the case in produced waters for instance, the amount of the waste disposal fee can be staggering.

The RCRA provides that various states can obtain EPA authorization to operate their own hazardous waste management programs in lieu of the federal RCRA program.[14] It may be of interest for California oil and gas producers to know that such authorization was effective as of August 1, 1992.[15]

Louisiana

By memorandum dated January 3, 1994, the Louisiana Commissioner of Conservation clarified the scope of the E&P exemption as applied to nonhazardous oilfield waste regulated under the Office of Conservation Statewide Order 29-B. That memorandum basically tracks the list of materials included in the EPA Federal Register publication, which appears in Appendix 7. It also incorporated materials discussed in the EPA's March 22, 1993 Federal Register publication, 58 Fed. Reg. 15284, which clarified the regulatory status of waste generated by crude oil reclaimers, service companies, gas plants, and pipelines.

In addition, the Office of Conservation administers the state's federally approved permitting program for saltwater injection wells.

Mid-Continent

The Oklahoma Corporation Commission (OCC) and Kansas Corporation Commission (KCC) are charged with regulating the disposal of produced waters. Kansas and Oklahoma generally require produced waters to be disposed of through underground injection. Regulations governing both commercial and non-commercial disposal wells are found in the OCC and KCC rules and regulations.

Rocky Mountain

The Rocky Mountain states all follow the RCRA exemption for oil and gas wastes in their state regulations. While most produced water is injected for enhanced recovery or in Class II injection wells (and in fact is the only disposal option in North Dakota[16]), in Wyoming and Utah significant amounts are disposed of through surface discharge.

CHAPTER 11 ENDNOTES

1. 42 USC §§ 6901–6992k.
2. This chapter is an adaptation of pertinent portions of article by Lori Wrotenbery, Director of Environmental Services for the Oil and Gas Division of the Railroad Commission of Texas, entitled Environmental Regulation in the Oil Field: Update on Railroad Commission Programs, State Bar of Texas Oil, Gas and Mineral Law Section Report, December, 1993.
3. Health and Safety Code §§ 25100, et seq.
4. Health and Safety Code § 25154.
5. Health and Safety Code § 25117.2.
6. Health and Safety Code § 25117(a)(1).
7. Health and Safety Code § 25141.
8. See, 14 California Code of Regulations §§ 1960, et seq.
9. Oil and Gas Journal, April 19, 1993 (Penwell Publishing Company), p. 32.
10. Oil and Gas Journal, April 19, 1993 (Penwell Publishing Company), p. 32.
11. 22 California Code of Regulations §§ 66264.270, et seq.
12. Health and Safety Code §§ 25174, et seq.
13. 22 California Code of Regulations §§ 66261.3(a)(2)(D), (E).
14. 42 USC § 6926.
15. 57 Fed. Reg. 32726 (July 23, 1992).
16. North Dakota Admin. Code 43–02–03–53.

12

NORM

Naturally occurring radioactive material (NORM) has long been recognized in association with oil and gas operations. NORM is found in produced waters, natural gas, drilling muds, filters, tank bottoms, resins, and scale in pipe. Although existing in varying degrees of concentration throughout the oil and gas industry, it is estimated that measurable quantities of NORM are found in only 15% of all production facilities.

It is not known what quantities of NORM pose dangers to human health and safety, especially when compared to normal background levels of radiation throughout the United States, including those resulting from living or working in brick structures or even from flying in airplanes. Nevertheless, certain states, such as Texas, have chosen to regulate the handling, storage, and disposal of NORM. To date, such regulations are by the states rather than by the federal government.

In Texas, NORM is regulated pursuant to the Texas Radiation Control Act.[1] Under this statute, the Texas Department of Health (TDH) regulates the possession, use, transfer, transport, and storage of NORM wastes and NORM-contaminated materials, such as drill pipe containing NORM scale. Only those materials for which radiation levels exceed the levels set forth in the "Texas Regulations for Control of Radiation" adopted by the TDH are subject to regulation, however.[2] These regulations also apply to those responsible for the handling of NORM materials, whether on a commercial for-hire basis, or merely as an incident to other operations, such as in oil and gas operations. The requirements governing the handling of NORM apply under the TDH regulations even though no specific requirements to measure or "survey" materials for radiation exist with respect to NORM materials in Texas, as contrasted with some other states.

Regulation of the disposal of NORM waste has been specifically given by statute to the Railroad Commission of Texas (RRC).[3] In consultation with the TDH, the RRC adopted its Oil and Gas Statewide Rule 94 effective February 1, 1995.[4] Under this rule, NORM waste with radiation levels exceeding specified levels are subject to specific regulation; disposal other than pursuant to the rule is prohibited.[5] Among other things, Rule 94: (i) allows disposal of NORM-contaminated produced waters through reinjection, (ii) allows disposal of NORM contaminated materials in plugged and abandoned wells, and (iii) specifically prohibits disposal of oil and gas NORM waste by spreading on public or private roads.

In adopting Rule 94, the RRC instructed its staff to continue studying these particular issues:

(i) the availability of field survey devices and methods;

(ii) the relationship between TDH exemption criteria and exposures associated with surface disposal; and

(iii) regulatory restrictions appropriate for disposal of buried pipes.

The RRC has stated its intention to amend the rules to reflect the results of these studies once they are complete.

Prior to the purchase of oil and gas properties, an investigation should be made to determine the extent to which NORM problems may exist. If problems exist, consideration should be given to the manner in which they will be managed, including the disposal of NORM-contaminated materials.

Louisiana

Louisiana has an extensive regulatory program covering NORM associated with oilfield operations, and was the first state in the country to address these issues. Louisiana's NORM program is primarily administered by the Department of Environmental Quality (DEQ).[6] Basically, those regulations establish a general licensing program for operators of sites where NORM has been detected either in equipment in levels exceeding 50 microroentgens per hour (including background), or in soil in levels exceeding 5 picocuries per gram (above background). Once a site is subject to the licensing program, worker protection requirements apply. In addition, the site cannot be transferred for "unrestricted use" (defined as the absence of any controls to protect the public from NORM exposure) unless remediation is performed to reduce soil NORM levels to not exceed 5 picocuries per gram above background (unless other levels are approved by the DEQ).

A recently instituted program allows certain NORM wastes generated during remediation projects to be disposed of at commercial nonhazardous oilfield waste disposal sites regulated by the Office of Conservation. Otherwise, transportation and disposal of NORM wastes are subject to special procedures set forth in the DEQ's regulations.

Parties considering purchasing Louisiana oilfield properties should confirm whether any NORM testing of the properties has been conducted (this testing, referred to as a "confirmatory survey," should have been done pursuant to the DEQ program), and whether the necessary labeling of any oilfield equipment has been done. In addition, if purchasing property that is represented to be free for "unrestricted use," the parties should ask for a written acknowledgment from the DEQ to that effect. Finally, parties considering remediation of NORM-contaminated sites should ensure that their cleanup contractors possess proper licensing authority from the DEQ, and that NORM waste will be treated and disposed of under proper procedures.

Rocky Mountain

While North Dakota has not adopted specific NORM regulations, the State Department of Health and Consolidated Laboratories (SDHCL) follows an informal policy that any material exceeding 5 picocuries/gram of total radium cannot be released for unrestricted use. Any persons having such materials in their possession are considered licensees and can only dispose of the materials in a manner approved by SDHCL.[7]

CHAPTER 12 ENDNOTES

1. Texas Health and Safety Code, Chapter 401.
2. The regulations are found principally in Parts 11, 21, 22, and 46 of the "Texas Regulations for Control of Radiation" adopted by the Texas Department of Health.
3. Texas Health and Safety Code, Section 401.415.
4. 16 TAC § 3.94.
5. The exemption level criteria are specified in the TDH's Regulations for Control of Radiation.
6. LAC 33:XV. 1401 at seq.
7. Conversation with Greg Krause, SDHCL, June 2, 1995.

13

Wetlands

Contrary to widespread misunderstanding, there is no legal prohibition against the acquisition or simple ownership of wetlands. Section 404 of the CWA (Clean Water Act), however, prohibits the discharge of materials onto wetlands without a permit from the U. S. Corps of Engineers (commonly referred to as a "Section 404 Permit").

Definition of Wetlands

In general, wetlands include swampy areas, mud flats and other areas that are generally, periodically, or only occasionally inundated by surface water. But the legal definition is set by the regulations of the Corps of Engineers, which specify technical tests involving measurement of hydrophytic vegetation, hydric soils, and hydrology. As a result, wetlands may be located not only in coastal or riparian areas, but also in areas far inland from the coast or well away from rivers. In cases where doubt exists, private engineering firms who specialize in wetlands evaluations should be consulted. In addition, an opinion should be obtained from the Corps of Engineers District Office with jurisdiction, since a consultant's opinion is not binding on the Corps.

Regulated Activities

The specific activities requlated by CWA Section 404 can be summarized as follows:

- Discharge of dredged material.
- Discharge of fill material for
  - construction.
  - impoundment of water.
  - site development.

  - causeways or roads.
  - shore protection devices such as breakwaters and seawalls.
- Discharge associated with excavation, including mechanized land clearing, ditching, or channelization. The Corps presently operates in accordance with what are called its Tulloch Rules, which are quite strict. Thus, as to land clearing, the Corps' present position is that using bulldozers, rakes, discs and backhoes is regulated, but not chain saw cutting trees above the soil surface.
- Pilings that have the functional effect of fill.

In short, most activities that can occur in drilling operations or in workover operations on an acquired oil and gas lease are regulated.

It should be noted, however, that only a portion of the land covered by a particular lease may actually be wetlands. In that case, if it is practical to limit operations to the portion of the lease that is not wetlands, the best solution is to avoid the necessity for a Section 404 Permit by restricting operations to the portion that is not wetlands.

It is of no concern to the buyer of oil and gas properties that the seller or an owner prior to the seller may have illegally conducted regulated activities without a Section 404 Permit. Caution must be exercised, however, if there are any notices of alleged violation or pending enforcement proceedings against a prior owner, since one sanction for enforcement is to require removal of unpermitted materials from the land. On the other hand, obtaining a Section 404 Permit is of concern to the buyer if continued operations involve a regulated activity, such as deposit of materials dredged to keep open waterway access to the lease, or if the buyer anticipates drilling or workover operations on the lease that will result in deposit of dredged or fill materials.

Section 404 Permit Process

General permits are promulgated either on a nationwide or regional basis within a particular Region of the Corps. Although general permits are issued for a fairly limited number of circumstances, they should always be considered. Qualifying under a general permit is more or less routine, but some general permits require advance notification to the Corps. One of the most frequently used general permits authorizes fill of up to 10 acres in "isolated" wetlands or "adjacent" wetlands above the headwaters of a stream.

In issuing individual Section 404 Permits, the Corps has authority to allow public comment after notice. However, as a practical matter, the Corps processes on the average over 15,000 applications per year and has required only about 50 public hearings per year.

Section 404 of the statute requires interagency coordination, and each application must be reviewed not only by the Corps but also by the EPA (Environmental Protection Agency), the U.S. Fish and Wildlife Service, and the National Marine Fisheries Service, as well as by appropriate state agencies. The Fish and Wildlife Service will routinely review an application for Endangered Species Act issues.

Strategies in the Permit Application Process

The object of the Corps is to preserve existing wetlands either by prohibiting regulated activities with an adverse impact or by mitigating the impact. In a sense, obtaining a Section 404 Permit is a negotiating process. You can expect the Corps' negotiating stance to move through three stages:

a. Change plans for development altogether to avoid adverse impact, if possible or
b. Modify plans for development to minimize impact or
c. Compensate by mitigation: create new wetlands or restore prior wetlands usually, but not necessarily, on a one-to-one basis.

There is a widespread general public misperception concerning the number of Section 404 Permits that are denied or granted. According to an EPA Wetlands Fact Sheet released in March, 1993, in 1992 the Corps reviewed 16,000 applications for individual permits and denied fewer than 400. Furthermore, approximately 80,000 additional activities were authorized by the Corps for general permits.

Louisiana

In Louisiana, the ability to locate and drill oil and gas wells in a coastal area is governed in part by the Louisiana Coastal Zone

Management Act.[1] Under this law, a coastal use permit is required for all activities having a direct and significant impact on coastal waters.[2] Thus, to site an oil well on Louisiana's coastal zone in an area that is also considered a "wetland," both a federal Section 404 Permit and a state coastal use permit (or drilling permit issued by the Office of Conservation) are required. However, a dual permit application scheme has been designed to ease this permitting process. In addition, Louisiana has issued a number of general coastal use permits, similar to the Corps of Engineers general permits, covering a variety of routine oil and gas activities.

Mid-Continent

Oklahoma law directs that an "Oklahoma Wetlands Management Strategy" be prepared by the Oklahoma Conservation Commission to define and identify all wetlands in the state and recommend measures to mitigate losses to wetlands while protecting the rights of landowners.

CHAPTER 13 ENDNOTES

1. La. R.S. 49:213.1 et seq.
2. La. R.S. 49:214.30.
3. 27A O.S. 3-2-108.

14

Endangered Species

The Endangered Species Act of 1973 (ESA) authorizes the U.S. Secretary of the Interior to list biological species of fish, wildlife, or plants determined to be threatened or endangered,[1] after following prescribed procedures. An endangered species is a species in danger of extinction,[2] and a threatened species is a species that is likely to become an endangered species within the foreseeable future.[3] There are currently more than 511 listed plants and more than 433 listed animals. Numerous additional species are under active consideration for listing. Additionally, at the time of listing or subsequently the Secretary of the Interior may designate critical habitat for a listed threatened or endangered species. Critical habitat, generally defined, is a geographical area occupied by a species that is essential for its conservation. A detailed discussion of the pertinent provisions of the ESA is found in Appendix 10.

Interagency Cooperation

The ESA can affect oil and gas operations in either of two ways. First, under the provision of the statute that all federal departments and agencies shall seek to conserve endangered species and threatened species, no federal agency can take any official action without consideration of the action's impact on threatened or endangered species. This is sometimes referred to as "interagency cooperation," and is expressly provided for in the ESA Section 7.[4] For example, the Corps of Engineers may not issue a Section 404 Wetlands Permit (as more particularly described in Chapter 13) nor may the Federal Energy Regulatory Commission issue a pipeline certification of convenience and necessity without, in each case, consulting with the U.S. Fish and Wildlife Service of the Department of the Interior concerning possible adverse effect on a threatened or endangered species or possible adverse effect on its critical habitat. The same rule applies to the granting of an oil and gas lease on federal lands.

Taking by a Private Person

Second, aside from the mandate for interagency cooperation in federal action, the ESA prohibits the taking of a listed species by any private person. Taking is defined broadly by statute to include any of several physical offenses, such as hunting, pursuing, killing, capturing, or harming.[5] By regulation, harming is defined to include significant habitat modification or degradation.[6] The ESA contemplates a permit for taking in special cases,[7] but the limitations on obtaining the permit and the length of time required to process the application effectively restrict its use to very exceptional circumstances. The most practical solution to potential taking problems is to contact the U.S. Fish and Wildlife Service in cases where doubt exists, and if a serious question exists, simply to move the site of operations.

State Endangered Species Acts

Most states now have some form of legislation similar to the ESA. Most of the states' acts either allow simply for the compilation of a list or regulate direct takings only. Few state acts provide any habitat protection provisions. The state acts frequently list species not considered threatened or endangered under the ESA. By way of illustration, in 1990 two pipeline maintenance workers were cited in Texas for taking timber rattlesnakes in violation of the Texas Endangered Species Act, although this species was not on the Department of the Interior's list.[8] Thus, consultation with the relevant state agency is indicated.

California

California has no endangered species law as such. However, as stated in Chapter 10, California does have many statutes prohibiting the endangerment by pollution of any form of plant or animal life.

Louisiana

In Louisiana, oil and gas operations are most likely to encounter state agency wildlife habitat concerns in the form of permit conditions covering seismic operations under a program administered by the Louisiana Department of Wildlife & Fisheries. In addition, oil and gas operations in the state's coastal zone, which would otherwise be subject to coastal use permits, may have to minimize impacts on wildlife habitat. While the latter program is administered by the Louisiana Department of Natural Resources, state commenting agencies include the Department of Wildlife & Fisheries. Specific examples of habitat concerns include possible disturbances of bird rookeries and eagle nests.

Mid-Continent

In Oklahoma, a list of endangered or threatened species protected by the Oklahoma Wildlife Conservation Code[9] can be obtained from the Department of Wildlife Conservation. The Kansas Department of Wildlife and Parks administers the Kansas Nongame and Endangered Species Conservation Act[10] and maintains a list of protected species.

Rocky Mountain

Rocky Mountain endangered species acts are the Wildlife Conservation Act[11] in New Mexico and the Nongame and Endangered Species Conservation Act[12] in Nebraska.

CHAPTER 14 ENDNOTES

1. 16 USC § 1532 (16).
2. 16 USC § 1532 (6).
3. 16 USC § 1532 (20).
4. 16 USC § 1536.
5. 16 USC § 1532 (19).
6. 50 CFR § 17.3.
7. 16 USC § 1539.
8. Personal conversation between the author and Peter G. Hirst, United Gas Pipeline Company, October, 1991.
9. 29 O.S. 2–101, et seq.
10. K.S.A. 32–801, et seq.
11. N.M. Stat. Ann. §§ 17–2–37 to –46.
12. Neb. Rev. Stat. §§ 37–430 to 37–438.

15

NPDES Permits

Control of water pollution under the Clean Water Act (CWA) is imposed primarily by prohibiting discharges of pollutants into surface waters, except as authorized by permit under the National Pollutant Discharge Elimination System (NPDES). A buyer of oil and gas properties ordinarily will not be liable for the past failure of its seller to obtain or comply with the NPDES permitting program. Nevertheless, a buyer who will acquire and operate oil and gas properties with surface water discharges will be held responsible for any noncompliance during the term of its ownership or operation. Thus, he should determine prior to closing whether the properties to be acquired include any discharges subject to NPDES permitting and whether those discharges are in compliance with the NPDES program requirements. A buyer should include in the due diligence a review of any existing permits held by the current operator or seller to determine whether existing discharges are in compliance with the permit conditions or are subject to any outstanding enforcement actions. In addition, he must provide a proper notice or application for the transfer of any required permit.

The CWA NPDES permitting program, which includes numerous General Permits issued for large geographic areas, is extensive and detailed. This chapter merely provides an overview of the NPDES permitting program and General Permits regulations and permitting rather than presenting a detailed analysis of all available permits.

The stated purpose of the CWA and its 1972 predecessor, the Federal Water Pollution Control Act, is to restore and maintain the chemical, physical, and biological integrity of the nation's waters. One of the principal programs of the CWA, as described in the CWA Section 402, is the NPDES permitting program. The statutory mechanism of the NPDES program is a broad prohibition on all discharges of pollutants into waters of the United States except under the specific permits or other authorization of the EPA (Environmental Protection Agency). The NPDES program applies to nearly all industrial wastewater treatment systems and untreated industrial discharges and, since the 1987 CWA amendments, also to discharges of storm water from many industrial facilities. Among the discharges that are prohibited without the authorization of an NPDES permit are those of oil and gas waste to surface water bodies.

Statutory **Basis** *of the* **NPDES** *Program*

The CWA Section 301 prohibits the "discharge of any pollutant by any person" without an NPDES permit or other authorization.[1] The applicability of the NPDES program depends primarily on the definition of these key terms:

- Discharge of a pollutant
- Pollutant
- Point Source
- Waters of the United States

The term **discharge of a pollutant,** is defined generally in the CWA as the addition of any pollutant to navigable waters from any point source.[2]

Pollutant, in accordance with the CWA definition, is any substance that, by its discharge into water, results in man-made or man-induced alteration of the chemical, physical, biological, and radiological integrity of the water.[3] A pollutant, consequently, includes any industrial waste discharged into water.[4] Produced water and other oil and gas wastes are industrial wastes and are subject to the NPDES program.

The term **point source,** is any discernable, confined, and discrete conveyance, including any pipe, ditch, conduit, well, discrete fissure, container, or vessel or other floating craft.[5] A point source can include a conduit that is nothing more than a ditch that has been graded and planted with grass, but there must be an element of human intervention that has the effect of concentrating the flow from the discharge in some manner. Wells, flow lines, separators, vacuum trucks, and fire walls can all be point sources, and discharges of produced water from such point sources are subject to CWA regulation, including NPDES permitting and enforcement.

The term **waters of the United States**, or "navigable waters," has been very broadly defined by both the EPA and the courts. It includes intermittent streams, mud flats, sand flats, wetlands, sloughs, prairie potholes, wet meadows, playa lakes, and natural ponds.[6] In fact, the discharge need not be directly to such waters; if a discharge leaves a point source and eventually makes its way to any one of these regulated waters, then it is subject to regulation under the NPDES program.

State NPDES Authority

The federal NPDES permitting program provides for delegation of NPDES permitting authority to states at the discretion of the EPA. Several major oil producing states have obtained such authorization,[7] but Louisiana, New Mexico, Oklahoma, and Texas have not been delegated the NPDES permitting authority. In those states, the EPA directly administers and issues NPDES permits for discharges, notwithstanding that surface water discharge authorizations or permits also may be required and issued under state law. Where the state that has not been delegated NPDES authority has its own permitting program, an operator must obtain authorization or a permit to discharge both from EPA and from the appropriate state administrative agency.

Even in those states with delegated authority, however, the EPA retains authority to review proposed permits and object to their terms or their issuance.[8] Furthermore, the EPA may take enforcement action against any discharges it believes to be in violation of provisions of the CWA or of a state-issued NPDES permit.[9]

Effluent Limitations Guidelines

NPDES permits usually contain effluent limitations for identified pollutants to establish the permissible level of discharge (if any) of specified substances. For example, oil and gas NPDES permits generally specify limits for concentrations of oil and grease in the permitted discharge. "Effluent Limitations Guidelines" are promulgated by EPA to establish technology-based effluent limitations that can be used by state or federal agencies as guidance in developing effluent limitations for NPDES discharge permits. The effluent guidelines for the "Oil and Gas Extraction Point Source

Category" of discharges are delineated in CFR Chapter 40, Part 435. This category is divided into five subcategories:

- Offshore
- Onshore
- Coastal
- Stripper
- Agricultural and Wildlife Water Use

Although the published Effluent Limitations Guidelines contain specific effluent limitations, for example, limits on daily maximum and monthly average concentrations in milligrams per liter of oil and grease for produced water discharges, they *do not* constitute authorization to discharge. They are intended for use only by the EPA or by states authorized to implement the NPDES program, as guidance in establishing the conditions that become part of NPDES permits. Regardless of the existence of published "effluent limitations guidelines," the CWA requires regulated discharges to be covered by an individual, group, or general NPDES permit. Discharges not covered by such permits are unlawful, even though they may be consistent with the Effluent Limitations Guidelines or with state permits, such as a Railroad Commission of Texas Permit to Discharge Produced Water.

NPDES Individual Permits and General Permits

The NPDES permit program contemplates the issuance of both individual permits, which are applicable to only one facility, and General NPDES Permits, which are applicable to a class of similar dischargers, such as the subcategories established in the Effluent Limitations Guidelines, within a discreet geographical area.[10] A General Permit is subject to rule-making procedures rather than the hearing procedures applicable to individual permits. Some of the General Permits that have been issued by the EPA are listed in Appendix 3. The EPA continues to issue new General Permits and modify existing ones. Over the past decade, it has declined to process and issue individual NPDES permits for oil and gas discharges, electing instead to administer the program through General Permits, some of which were not issued until 1995. Recently, however, the EPA has reached an agreement with the

Railroad Commission of Texas whereby individual NPDES permit applications for oil and gas facilities that are submitted to EPA Region 6 can be processed by the Railroad Commission and issued by the EPA in a timely manner.

NPDES Storm Water Discharge Permitting Program

In 1987, Congress amended the CWA to require the EPA to establish a permitting program for point source discharges of storm water. The purpose of the program is to control pollutants contained in storm water runoff. To implement this statutory extension of the NPDES program, the EPA developed storm water discharge permit application requirements, which are found in 40 CFR Part 122.[11] The storm water discharge regulations cover discharges of "storm water associated with industrial activity," which is defined to include discharges of storm water from oil and gas exploration and production activities.[12]

The operator of an oil or gas production, processing, or treating facility is not required to submit an NPDES storm water discharge permit application unless the facility has had a discharge of storm water resulting in the discharge of a *reportable quantity* of either oil or a hazardous substance since November 16, 1987 or has such a release in the future.[13]

The term "reportable quantity," is defined as either: (a) an amount of oil or a hazardous substance that violates applicable state water quality standards or (b) an amount that causes a film or sheen upon or discoloration of the surface of the water or adjoining shorelines, or that causes the deposit of a sludge or emulsion. Under the statute, the existence of a reportable quantity of oil or hazardous substance alone is not grounds for requiring an NPDES storm water permit; a permit is required only if the storm water runoff has come into physical contact with specified materials at the site of the operation.[14] These materials are overburden, raw material, intermediate products, finished products, by-products, or waste products located on the site.[15] Administratively, however, the EPA has taken the position that this statutory specification of such physical contact does not restrain it from requiring a permit following discharge of a reportable quantity. The EPA's regulations presently require a permit in such cases regardless of the facts pertaining to physical contact.[16]

In addition to the storm water regulations, the EPA has issued two General Permits that apply to storm water discharges in certain states, as discussed below.

General NPDES Storm Water Permit for Industrial Activity

On September 9, 1992, the EPA issued a General Permit for NPDES storm water discharges, which applies to eleven states, including Louisiana, New Mexico, Oklahoma, South Dakota, and Texas.[17] The operator seeking to be covered by this General Permit must file a Notice of Intent. Oil and gas facilities that become subject to the storm water discharge permitting requirements as a result of a discharge of a reportable quantity of oil or hazardous substances must submit a Notice of Intent within 14 days after the operator first has knowledge of the release. If the operator becomes no longer responsible for the discharges from a particular facility, he should file a Notice of Termination, thereby shifting responsibility for the discharge from himself to a new operator.

An operator who files a Notice of Intent agrees that it will comply with the permit conditions imposed by the General Permit. The conditions of this General Permit include a prohibition against commingling non-storm water discharges and a requirement that the operator prepare and implement a storm water pollution prevention plan.

Criminal and civil penalties for violations of permit conditions are the same as penalties under the CWA for discharging without a permit.

General NPDES Permit for Storm Water Runoff from Construction Activities

Although the CWA does not distinguish between "industrial" and "construction" activities, the EPA has published a separate storm water General Permit for construction activities that applies in ten states, including Louisiana, New Mexico, Oklahoma, South Dakota, and Texas.[18] This General Permit applies to construction activity involving clearing, grading, and excavation that results in the disturbance of five acres or more of total land area. The Construction General Permit applies to linear projects, such as pipeline installation or maintenance, when the overall project will result in the disturbance of five acres or more. In a recent case, the Ninth Circuit Court of Appeals struck down the exclusion of tracts less than five acres as being arbitrary and capricious.[19] The EPA, however, still has not issued a General Permit that would apply to construction projects on less than five acres.

The operator of the construction activity seeking to be covered by this General Permit must submit a Notice of Intent to be covered. The operator can file a Notice of Termination upon completion of the construction project or upon sale or transfer of the project to a new operator.

An operator who files a Notice of Intent agrees that it will comply with the conditions imposed by the Construction General Permit. Similar to the General Permit for Industrial Activities, the Construction General Permit conditions include a prohibition against commingling non-storm water discharges, and requires preparation and implementation of a storm water pollution prevention plan.

Louisiana

Although Louisiana has not been delegated NPDES authority, the state is in the process of applying for the authority. In the meantime, the Louisiana Department of Environmental Quality administers a separate and parallel water discharge permitting program pursuant to the Louisiana Control Law.[20]

Rocky Mountain

The Rocky Mountain states include a large number of Indian reservations that contain significant amounts of oil and gas properties. These include the Southern Ute and Ute Mountain Ute reservations in Colorado, the Jicarilla Apache reservation in New Mexico, the Navajo reservation in New Mexico and Utah, the Three Affiliated Tribes reservation in North Dakota, the Northern Ute reservation in Utah, and the Wind River reservation in Wyoming.

Generally speaking, states do not have jurisdiction over activities on tribal land and thus state environmental laws have no effect there. However, federal environmental laws do apply on tribal lands, and under the Clean Water Act, the Clean Air Act, the Safe Drinking Water Act and CERCLA, the EPA is expressly permitted to treat tribal governments in the same way as state governments. Accordingly, under those acts, the tribal government could receive implementing authority in the same fashion as a state. To date, the tribal governments generally have not sought implementing authority and thus the EPA is responsible for implementing these programs on tribal lands.

The tribal governments also have the authority to adopt and enforce their own environmental regulations. There appears to be increasing interest among the tribes to take action in this area, and in particular the Navajo Nation and the Jicarilla Apaches have adopted a significant number of tribal environmental regulations.

Thus, purchasers of oil and gas properties located on tribal lands should be sure to carefully investigate the tribal environmental regulations that apply to those properties, since they may be subject to a very different regulatory scheme than properties off the reservation. In addition, enforcement of these tribal regulations is in tribal court, not state court, which should give any purchaser extra incentive to carefully review the compliance status of the properties.

CHAPTER 15 ENDNOTES

1. 33 USC § 1311(a).
2. 33 USC § 1362(12).
3. 33 USC § 1362(19).
4. 33 USC § 1362(6) (definition of pollutant), (12) (definition of discharge of a pollutant), and (16) (definition of discharge).
5. 33 USC § 1362(14).
6. 40 CFR § 122.2.
7. Oil producing states which have obtained such EPA delegation of the NPDES permitting authority include California, Colorado, Kansas, South Dakota, Utah and Wyoming.
8. 40 CFR § 123.44.
9. 33 USC § 1319(a).
10. 40 CFR § 122.28.
11. 40 CFR § 122.26.
12. 40 CFR § 122.26(b)(14)(iii).
13. 40 CFR §§ 122.26(b)(14)(ii)–(iii), 122.26(c)(1)(iii)(A)–(B).
14. 33 USC § 1342(1)(2).
15. 40 CFR § 122.26(c)(1)(iv).
16. 55 Fed. Reg. 48032.
17. 57 Fed. Reg. 41236 (Sept. 9, 1992). This general permit applies also to Alaska, Arizona, Florida, Idaho, Maine, and New Hampshire.
18. 57 Fed. Reg. 41176 (Sept. 9, 1992). This general permit applies also to Alabama, Alaska, Idaho, Maine, and New Hampshire.
19. NRDC v. EPA, 966 F.2d 1292 (9th Cir. 1992).
20. La. R.S. 30:2071 et. seq.

16

SPCC *Plans*

The requirements of the law for Spill Prevention Control, and Countermeasure (SPCC) plans as discussed in this chapter are prospective. Nevertheless, in any acquisition transaction, the buyer must evaluate the seller's SPCC plans that are in place in contemplation of inadequate or non-conforming plans. These should be corrected by the seller as a condition of the transaction or else the buyer must anticipate the cost and the EPA (Environmental Protection Agency) involvement of bringing the plans into compliance.

Federal law requires SPCC plans for oil and gas facilities (either onshore or offshore) engaged in either drilling or producing, gathering or storing, processing or refining, or transferring, distributing, or consuming oil and oil products.[1]

Applicable Statutes *and* Regulations

Since the early 1970s, the Clean Water Act has required the EPA to establish procedures, methods, and equipment and other requirements for equipment to prevent the discharge of oil and hazardous substances from vessels, onshore facilities, and offshore facilities, and to contain such discharges.[2]

In response, the EPA developed its SPCC regulations. The current version of the regulations is found in Part 112 of Chapter 40 of the Code of Federal Regulations. Part 112 applies only to facilities that, because of their location, "could reasonably be expected to *discharge oil* in *harmful quantities* . . . into or upon *navigable waters* of the United States or adjoining shorelines."[3] The application of the Part 112 SPCC plan requirements depends on the definitions of these terms:

- harmful quantities
- discharge

- navigable waters
- oil

The term **harmful quantities,** by reference to Section 110.3 of Chapter 40 of the Code of Federal Regulations, is defined as that quantity of oil that will:

> *cause a film of sheen upon or discoloration of the surface of the water or adjoining shorelines or cause a sludge or emulsion to be deposited beneath the surface of the water or upon adjoining shorelines.*[4]

This is widely known as the "oil sheen test."

The term **discharge** under Part 112 is not just a catastrophic event but also includes any spilling, leaking, pumping, pouring, emitting, emptying, or dumping.[5]

Navigable waters has been interpreted very broadly by both the courts and the EPA to include lakes, rivers, streams, wetlands, and even dry arroyos.[6] Part 112 expressly states that the determination of applicability must be based solely upon a consideration of the proximity of navigable waters to the facility, taking into consideration geographical factors such as land contour and drainage. The consideration *must exclude* all manmade features, such as firewalls, dikes, containment structures, and automatic shutoff valves designed to prevent a discharge of oil in harmful quantities.[7] Thus, when determining whether a facility is subject to the Part 112 SPCC plan requirements, the operator must evaluate its potential to discharge based primarily on tank capacity, geography, and drainage, disregarding any artificial structures in place to contain leaks and spills.

Oil is defined very broadly as:

> *Oil of any kind or in any form, including, but not limited to petroleum, fuel oil, sludge, oil refuse and oil mixed with wastes other than dredged spoil.*[8]

Although the EPA has not issued any formal guidance on the definition of oil, it has made informal guidance available. The EPA has prepared a draft list of oils and non-oils, which is contained in Appendix 4. The EPA includes such substances as mineral spirits, refinery petroleum waste, and naphtha in the definition of oil, while such substances as ethylene glycol, benzene, and natural gas condensate are not considered to be oils. The most conservative course of action in determining which substances are regulated as oil is to request a written opinion from the EPA for the substance in question.

Despite the broad reach of the SPCC plan requirements under these definitions, a facility will not be subject to the SPCC requirements unless it is a facility with either an: (i) aggregate above ground storage capacity of more than 1320 *gallons* (not barrels), or less than 1320 gallons if any single container has a capacity greater than 660 gallons; or (ii) underground storage capacity of more than 42,000 gallons.[9] Transportation related facilities are regulated by the Department of Transportation, and therefore are not covered by Part 112 SPCC plan regulations. The EPA estimated that in 1994 there were over 450,000 facilities subject to the SPCC plan regulations, and anticipated the number will increase by over 4,500 facilities per year.[10]

SPCC *Plan Requirements*

In stark contrast to programs requiring spill response plans, the SPCC plan program focuses on prevention of, rather than reaction to, sudden discharges or catastrophic events. In general, owners or operators of facilities subject to the Part 112 SPCC plan requirements must prepare an SPCC plan, maintain it, and comply with prescribed record-keeping requirements. They must also revise the plan in the event of facility modifications, submit it to the EPA in the event of certain discharges of oil, and conduct a review of it every three years. A detailed discussion of specific requirements for preparation and implementation of the SPCC plan follows.

Plan Preparation and Implementation. Owners and operators of facilities subject to SPCC plan requirements must prepare a written plan within six months after the date a facility becomes operational. The plan must be fully implemented as soon as possible, but no more than one year after operations commence.[11] The EPA may authorize extensions of time for both preparation and implementation of the plan requirements where conditions, such as nonavailability of required equipment or construction delays, are both outside the control of and not the fault of the owner or operator.[12] The owner or operator seeking an extension must submit the request in writing to the EPA Regional Administrator. The request must be accompanied by a complete copy of the SPCC plan (if available) and a complete discussion of the delay, its cause, mitigating actions, and a proposed time schedule for implementation.[13] If an extension is granted, it will apply only to the specific aspect of the SPCC plan that has been affected by the unavoidable delay.[14]

Owners and operators of mobile or portable facilities, such as onshore drilling rigs or offshore barge-mounted rigs, must either comply with the requirements to prepare a specific plan for each location or may prepare a general plan applicable to any site. If the general plan option is selected, then the SPCC plan requirements are to be implemented upon installation at the new location and no grace period is available.[15]

Every SPCC plan must be reviewed and certified by a Professional Engineer. The owner or operator is responsible for maintaining a complete copy of the plan at any facility that is attended at least eight hours per day during the workweek. For unattended facilities, such as production locations and tank farms, a copy of the current plan must be available at the nearest field office so that it is available to EPA inspectors during normal working hours.[16]

Reporting, Updating, and Record-Keeping Requirements. Once the owner or operator has fully implemented a required SPCC plan, additional spill reporting requirements are triggered. If the facility discharges more than 1,000 gallons of oil into "navigable waters" in any one spill event or has two or more discharges of "harmful quantities," *i.e.*, quantities that produce a sheen, within any twelve-month period, then *in addition to any other state or federal reporting requirements,* the owner or operator must submit to its EPA regional administrator a detailed report within sixty days.[17] The contents of the required report are found in Appendix 5. A copy of the report must be provided to the state agency with jurisdiction over oil- and gas-related water pollution control activities. The operator should contact the EPA regional office or the state agency that regulates oil and gas activities to verify which state agency or agencies should receive this report. In response, the state agency may conduct an investigation and make recommendations to the EPA regarding the procedures, methods, and equipment that may be necessary to prevent and contain a spill from that facility. The EPA, on its own initiative or in response to the state's comments, may require an amendment to the plan.[18] Notice of the amendment requirements and an opportunity to respond are required by the rules, and the owner or operator may appeal the decision to the EPA.[19]

Owners and operators of facilities subject to SPCC plan requirements must amend the plan when necessary. At least once every three years, the plan must be reviewed and evaluated.[20] If the three-year periodic evaluation shows that more effective prevention and control technology exists that will significantly reduce the likelihood of a spill and if the technology has been proven in the field, then the plan must be amended within six months to incorporate the new technology.[21] Additionally, the plan must be amended whenever the owner or operator changes the facility design, construction, operation, or maintenance in a manner that materially affects the potential to discharge.[22] All amendments must be fully implemented within six months and must be certified by a Professional Engineer.[23]

Training and Security. Owners and operators are responsible for personnel training, which includes the proper operation and maintenance of the equipment to prevent the discharges of oil. In addition, each facility must have a designated person who is responsible for oil spill prevention. Owners and operators are required to develop procedures to assure that new personnel are familiar with the SPCC plan. Periodic briefings should be held as part of these procedures.[24]

At all non-production facilities, security must be provided as reasonably necessary to prevent spills resulting from vandalism or accidents. The plant must be fully fenced and entrance gates must be locked or guarded when the plant is unattended.[25] Valves, pipeline loading connections, and starter controls on pumps must be turned off and secured when not in use or otherwise needed for operations.[26] Additionally, the facility must be lighted in such a manner as to reveal spills occurring during hours of darkness and to prevent spills through acts of vandalism.[27]

Technical Guidelines for SPCC Plans

Section 112.7 of Part 112 contains a detailed discussion of technical guidelines for preparing SPCC plans. Generally, the plan must conform to "good engineering practices." As with any effective environmental program, the SPCC plan requirements must have full approval and support through the highest levels of management.[28]

All SPCC plans should follow the sequence outlined in section 112.7. The guidelines set forth the following elements of a required plan:

- A written description of each spill event within 12 months prior to January 10, 1974 (the effective date of the Part 112 regulations), including a written description of each spill, corrective action taken, and subsequent preventive measures.[29]
- A prediction of the direction, rate of flow, and total quantity of oil that could be discharged from equipment that has a reasonable potential for a major type of failure. Although professional judgment must be exercised in determining which equipment should be included, the plan should include at a minimum any tank with a potential to overflow, rupture, or leak.[30]
- For each potential discharge described above, list appropriate containment and/or diversionary structures and equipment available to prevent discharges of oil from reaching a water course. Minimum requirements include:
  - For onshore facilities; sufficiently impervious dikes or berms, curbing, culverts, or other drainage systems, booms or other physical barriers or sorbent materials and spill diversion or retention ponds.
  - For offshore facilities, curbing, drip pans and sumps, or other types of collection systems.[31]

  Any facilities not fully operational must be discussed separately to include dates of completion and start-up.[32]
- If the installation of the diversionary structures or equipment listed above is not practicable, then the owner or operator must provide a clear demonstration of the impracticability and include the following:
  - An oil spill contingency plan consistent with the requirements of Part 109 of Chapter 40 of the Code of Federal Regulations and a written commitment of the manpower, equipment, and materials necessary to "expeditiously control and remove any harmful quantity of oil discharged."[33]

- A complete discussion of conformance with the following specific guidelines:
  - For onshore facilities other than production facilities, e.g., bulk storage facilities and transfer operations,[34] the plan must include a discussion of drainage systems that meet specific engineering requirements.[35]
  - All onshore non-production bulk storage facilities must meet material and construction requirements to assure compatibility of temperature and pressure conditions with the materials stored.[36] They must provide for a means of secondary containment for the "entire contents of the largest single tank plus sufficient freeboard to allow for precipitation."[37] In addition, they must meet specific engineering requirements for drainage of spills and rain water and for cathodic protection and tank integrity including testing, inspection requirements, fail-safe engineered installation, observation of plant effluents, prompt correction of visible oil leaks, and consideration of the positioning of mobile or portable oil storage tanks to minimize the potential for discharges.[38]
  - For facility transfer operations (excluding production facilities), the plan must comply with specific engineering requirements for buried piping installations, out-of-service pipelines, pipe supports, and above ground valves and pipelines. These include cathodic protection, other abrasion and corrosion protection, regular inspections, testing, and warning signs, particularly for vehicle traffic.[39]
  - For facility tank car and tank truck loading or unloading racks, the plan must describe minimum engineering requirements for the facility that include containment designed to hold at least the maximum capacity of any single compartment of a tank truck or car, safety design features such as physical barriers of warning lights, and systematic inspection of vehicle drains and outlets.[40]
  - For oil production facilities, which include "all wells, flowlines, separation equipment, storage facilities, gathering lines, and auxiliary non-transportation-related equipment and facilities in a single geographical oil or gas

field operated by a single operator,"[41] the plan must provide for closure of dike drains and for inspection of any rainwater prior to discharge. It must have provisions for removal of any accumulated oil in diked areas, drainage ditches, road ditches, and sumps or similar facilities at regular intervals.[42]

- Bulk storage tanks at oil production facilities should meet guidelines similar to those for bulk storage tanks discussed above except that regularly scheduled periodic visual inspection may substitute for cathodic protection and integrity testing.[43]
- Facility transfer operations should comply with the guidelines for periodic scheduled examination of various valve and pipeline components and must have a maintenance program, including inspections, corrosion protection, replacement schedules, and adequate record keeping for each individual facility.[44]
- Oil drilling and workover facilities should meet guidelines requiring catchment basins or diversion structures sufficient to contain spills of oil. In addition, a blowout prevention assembly and a well control system must be installed before drilling below any casing string or during workover operations.[45]
- Offshore oil drilling, production, or workover facilities must provide for oil drainage collection equipment to prevent and control small oil spillage around pumps, valves, flanges, and other equipment. Drains and sumps should be used where practicable. However, if they are not practicable, oil contained in collection equipment must be removed as often as necessary to prevent overflow.[46] Storage tanks should be equipped with appropriate protection devices, including level sensing devices, high and low pressure devices if it is a pressure tank, and corrosion protection.[47] The facility also must have a written procedure for inspection and testing of equipment on a periodic basis.[48] The plan must include consideration of well-control requirements,[49] pipelines, and transfer operations,[50] and provide for coordination with service companies and subcontractors.[51]

The engineer reviewing or preparing the SPCC plan should include a discussion of each specific requirement of Part 112 applicable to the facility.

Civil Penalties

Owners or operators subject to the SPCC plan requirements who fail or refuse to comply with regulatory requirements, including preparing, implementing, maintaining, and amending an adequate plan, are subject to civil penalties.[52] The EPA can assess civil penalties of up to $5,000 for each day of a continuing violation.[53] In determining the penalty amount, the EPA will consider the seriousness of the violation and the operator's demonstrated good faith efforts to bring the facility into compliance as soon as possible after receiving notice of the violation.[54] Each violation is subject to the operator receiving a notice of violation as well as an opportunity for a hearing.[55]

Available Sample Plans

The American Petroleum Institute has published its API Bulletin D16, "Suggested Procedure for Development of Spill Prevention Control and Countermeasure Plans," which contains suggested model forms that can be used to comply with existing SPCC plan requirements. In addition, EPA Region 6 has published a brochure titled "Information on SPCC Plans," which includes a sample plan.

Inspections

A written record of the inspection must be signed by the appropriate supervisor or inspector, made a part of the plan, and maintained for a period of three years.[56] The SPCC plan should provide for equipment testing and inspections on a scheduled periodic basis "commensurate with the complexity, conditions, and circumstances" of the facility.[57]

In preparing for the possibility of a compliance inspection, a facility owner or operator should conduct his own inspections, and review health and safety requirements with the employees to ensure that the facilities are being operated in compliance with all applicable rules and regulations. In addition, files and records should be reviewed to ensure that all recordkeeping requirements are being met. If the owner has any questions regarding privileged or confidential documents that may be subject to review by inspectors, then he should contact his attorney to discuss these issues prior to the inspection. It can be quite helpful in anticipation of any type of inspection to obtain a copy of the inspection form or internal inspection guidance that the agency has developed. In response to a federal Freedom of Information Act request, EPA Region 6 provided a copy of its "SPCC Compliance Inspection Report" form. That form is included for reference as Appendix 6. In preparing for an SPCC compliance inspection, a facility should take into account the eleven potential "deficiencies" listed in the form, and take steps to ensure that, at the very least, its facility does not exhibit any of these.

During an inspection, the owner or operator should make every effort to have a representative accompany the inspectors and keep them in one group. The owner should assign one manager or upper level supervisor to be present during the entire inspection and to take duplicate photos or video tapes if possible. That individual should ask questions about the concerns, if any, expressed by inspectors during the visit and should request copies of inspection reports or check-sheets filled out during the inspection. If possible, management should meet with agency representatives for a "debriefing" immediately following completion of the inspection. Such a meeting fosters good relations between the agency and the facility owner and has the potential to avoid enforcement actions by answering questions and providing supporting documentation. In general, the owner should make every effort, for both safety and liability reasons, to control the inspection while cooperating with agency officials to the greatest extent possible.

Proposed Further Regulation

Although the Oil Pollution Act of 1990 (OPA) included many amendments to the Clean Water Act, it did not amend the specific language of Section 311 of the Clean Water Act affecting the SPCC program requirements. In response to particular requirements of the OPA, the EPA has promulgated amendments to Part 112 regarding the OPA response plans.[58] On February 17, 1993 the EPA proposed certain changes to the Part 112 SPCC plan regulations in conjunction with proposed rules on facility response plans required by the OPA. Although the EPA did not adopt the proposed SPCC plan regulatory changes upon promulgation of the final response plan regulations, the issue addressed by the proposal may appear in future SPCC regulation amendments. They included authority for the EPA to amend the SPCC plan when it does not meet regulatory requirements and to regulate exempted facilities on a case-by-case basis.[59] Ultimately the OPA regulatory changes left the SPCC plan requirements untouched.

Prior to the OPA regulations, the EPA had established an SPCC Task Force that recommended certain changes to the EPA's SPCC regulations alone. In response to those recommendations, the EPA published proposed rules on October 22, 1991, but that proposal has not been finalized and is not currently in effect. Those proposed rules leave the basic structure of the SPCC plan requirements intact, but would impose additional regulatory burdens on owners and operators. The proposed regulations reflect the EPA's intent that the SPCC plan requirements be mandatory, absent a showing of "equivalent alternate protection."[60] The proposed rules substitute the words "requirements" and "shall" for the words "guidelines" and "should" in the sections governing the basic plan requirements and the proposed rules elaborate on specific requirements. For example, under the proposed rules, the plan must be certified by a Professional Engineer, as presently required, with the added limitation that the engineer cannot be either an employee of the facility or an individual who has any direct financial interest in the facility.[61] Dikes and berms must be impervious to oil for 72 hours, instead of "sufficiently impervious;"[62] integrity tests of tanks must be conducted every five years instead of every ten, and must be conducted annually where the facility has no secondary containment;[63] and valves and piping in transfer facilities must be inspected monthly instead of "periodically on a scheduled basis."[64] Lastly, spills and accumulations of oil must be removed within 72 hours, instead of simply "removed" with no time period indicated in the rules.[65]

Conclusion

The EPA requires a facility owner and/or operator to prepare an SPCC plan for all non-transportation-related oilhandling facilities with above ground storage capacity of over 1,320 gallons, where any single container holds in excess of 660 gallons, and below ground storage capacity over 42,000 gallons. Compliance requires preparation of a plan that conforms with the elements expressly stated in Part 112 of Chapter 40 of the Code of Federal Regulation, including certification by a registered professional engineer. A copy of the plan must be kept at the facility or nearest field office and the EPA has the right to conduct unannounced inspections to view the plan and enforce compliance with it. Violation of SPCC plan requirements is punishable by an administrative penalty of up to $5,000 per day.

California

Under California regulations, oil spill contingency plans are required both for onshore installations[66] and for offshore installations.[67]

Louisiana

The Louisiana Department of Environmental Quality administers a spill prevention and control plan under the state's parallel water discharge permitting program.[68]

Mid-Continent

In addition to the federal SPCC mandates, Oklahoma and Kansas have specific reporting requirements for spills and leaks of deleterious substances used in the operation of oil and gas wells that cause contamination of surface waters or groundwaters. The buyer should contact the Oklahoma Corporation Commission, the Kansas Corporation Commission, and the Kansas Department of Health and Environmental to ascertain the details of each states' reporting requirements.

CHAPTER 16 ENDNOTES

1. 40 CFR § 112.1(b).
2. 33 USC § 1321(j)(1)(C).
3. 40 CFR § 112.1(b).
4. 40 CFR § 110.3(b).
5. 40 CFR § 112.2 (definition of discharge).
6. 40 CFR §§ 110.1 (definition of navigable waters) and 112.2 (definition of navigable waters); United States v. Riverside Bayview Homes, Inc., 474 U.S. 121, 131–135 (1985); Leslie Salt Co. v. United States, 896 F.2d 354, (9th Cir. 1990); Quivira Mining Co. v. EPA, 756 F.2d 126 (10th Cir. 1985); U.S. v. Byrd, 609 F.2d 1204 (7th Cir. 1979).
7. 40 CFR § 112.1(d)(1)(i).
8. 40 CFR § 112.2 (definition of oil).
9. 40 CFR § 112.1(d)(2).

10. EPA Notice, 59 Fed. Reg. 34,612 (1994).
11. 40 CFR § 112.3(a)–(b).
12. 40 CFR § 112.3(f).
13. 40 CFR § 112.3(f)(2).
14. 40 CFR § 112.3(f)(3).
15. 40 CFR § 112.3(c).
16. 40 CFR § 112.3(e).
17. 40 CFR § 112.4(a).
18. 40 CFR § 112.4(d).
19. 40 CFR § 112.4(e)–(f).
20. 40 CFR § 112.5(b).
21. 40 CFR § 112.5(b).
22. 40 CFR § 112.5(a).
23. 40 CFR § 112.5(a)–(c).
24. 40 CFR § 112.7(e)(10)(iii).
25. 40 CFR § 112.7(e)(9)(i).
26. 40 CFR § 112.7(e)(9)(ii)–(iv).
27. 40 CFR § 112.7(e)(9).
28. See 40 CFR § 112.7 (preamble).
29. 40 CFR § 112.7(a).
30. 40 CFR § 112.7(b).
31. 40 CFR § 112.7(c).
32. 40 CFR § 112.7 (preamble).
33. 40 CFR § 112.7(d).
34. Production facilities are defined to include "all wells, flow lines, separation equipment, storage facilities, gathering lines, and auxiliary non-transportation related equipment and facilities in a single geographical oil or gas field operated by a single operator." 40 CFR § 112.7(e)(5) (definition of oil production facilities [onshore]).
35. 40 CFR § 112.7(e)(1).
36. 40 CFR § 112.7(e)(2)(i).
37. Id. § 112.7(e)(2)(ii).
38. Id. § 112.7(e)(2)(ii)–(xi).
39. 40 CFR § 112.7(e)(3).
40. 40 CFR § 112.7(e)(4).
41. 40 CFR § 112.7(e)(5)(i).
42. 40 CFR § 112.7(e)(5)(ii).
43. 40 CFR § 112.7(e)(5)(iii).
44. 40 CFR § 112.7(e)(5)(iv).
45. 40 CFR § 112.7(e)(6).
46. 40 CFR § 112.7(e)(7)(ii).
47. 40 CFR § 112.7(e)(7)(v)–(vii).
48. 40 CFR § 112.7(e)(7)(viii)–(ix).
49. 40 CFR § 112.7(e)(7)(xi).
50. 40 CFR § 112.7(e)(7)(xiv)–(xviii).
51. 40 CFR § 112.7(e)(7)(xiii).
52. 40 CFR §§ 112.6, 114.1, & 114.2.
53. 40 CFR §§ 112.6(b) & 114.1(b).

54. 40 CFR § 114.3.
55. 40 CFR §§ 114.4 & 114.5.
56. 40 CFR § 112.7(e)(8).
57. See 40 CFR § 112.7(e)(7)(ix).
58. 40 CFR § 112.20 – 112.21.
59. 58 Fed. Reg. 8841–8844 (1993).
60. 56 Fed. Reg. 54,620 (1991).
61. *Id.* at 54,618.
62. *Id.* at 54,621.
63. *Id.* at 54,621.
64. *Id.* at 54,624.
65. *Id.* at 54,637.
66. 14 California Code of Regulations § 1722.
67. 14 California Code of Regulations § 1743(d).
68. LAC 33:IX.901 et seq.

17

Clean Air Act

It is sometimes thought that environmental regulation under the Clean Air Act (CAA) does not significantly affect exploration and production operations. As discussed in this chapter, there are, however, CAA concerns that must be considered in the acquisition of producing properties. Additionally, the buyer must consider the possible problems from pollutant emissions that can arise under the common law of nuisance, outside of the statute.

The buyer of oil and gas properties is not responsible for the seller's failure to comply with the statute and regulations in operations prior to the acquisition. Nevertheless, the buyer must assess the seller's compliance, in part to be sure that continued operations by the buyer do not inadvertently continue a failure by the seller to obtain the requisite permits or to comply with the requirements of permits that have been issued.

The CAA, together with administrative regulations under it, is perhaps the most complicated and detailed of all federal environmental regulations. A general overview of the CAA scheme of regulation is necessary as a framework for discussing its application to exploration and production. The text of this chapter is intended to present only a synopsis of the more salient provisions of the CAA; a more detailed discussion is contained in Appendix 9.

CAA Overview

Under the CAA, air pollutants are categorized either as *criteria air pollutants,* as specified in the National Ambient Air Quality Standards (NAAQS), or as *hazardous air pollutants (HAPs).* Criteria air pollutants include precursors to criteria air pollutants. For instance, volatile organic compounds (VOCs) are a precursor to ozone, which itself is a criteria air pollutant. Criteria air pollutants include nitrogen dioxide (NO_2), carbon monoxide (CO), sulfur dioxide (SO_2), lead, particulate matter less than 10 microns in diameter

(PM–10), and ozone (O_3). In exploration and production, these are found in the products of combustion; in the case of ozone, they are formed by photochemical reactions involving NO_x and VOCs. HAPs come from either combustion processes, a gas stream, or from chemicals used in the processing of gas, such as diethanolamine (DEA).

The CAA dictates different regulatory systems for the criteria air pollutants and the HAPs. For criteria air pollutants, each state is divided into one or more air quality control regions (AQCRs). Each AQCR is then classified as attainment, nonattainment, or unclassified (which is treated as if it were attainment) for each of the criteria air pollutants. Based upon this classification, particular requirements are imposed. Before a facility can be built or modified, a new source review must be performed. The new source review requirements for an attainment AQCR are different from those for a nonattainment AQCR. For an attainment area, the requirements are set forth in the prevention of significant deterioration program (PSD). Nonattainment area requirements are provided in the nonattainment program.

HAPs are regulated under CAA Section 112. A compound can be regulated as a HAP and also as a precursor to a criteria pollutant. This explains the reason that volatile organic compounds (VOCs) may be regulated both as criteria air pollutants or as HAPs. HAPs such as n-hexane and benzene are also VOCs.

Permits

The CAA regulatory system uses permits to control pollutant emissions in order to protect or improve air quality. The permit establishes legal requirements for the operation of a source or facility. These constitute the law for that particular facility. Hence, the buyer must first determine if the facility needs a permit and then whether it has been put in place. The buyer then should review any permits to determine that the facility is in compliance with them. These various types of permits must be considered:

- Preconstruction permit for attainment areas
- Preconstruction permit for a nonattainment area
- Operating permit (now required under the CAA Amendments of 1990)
- Standard or general permit for the facility

The permit establishes the standards of performance for the facility. If the permit provisions are violated, then the facility is in violation of the CAA. The facility must also comply with the provisions of the State Implementation Plan (SIP), which are generally incorporated either explicitly or by reference in the permit. The SIP must provide for the overall permit program and for measures that will ensure that air quality attains and maintains the NAAQS. It should be reviewed for its impact on the facility if any of its provisions are incorporated by reference into any permit that affects the facility. Future operating permits will explicitly incorporate such requirements.

Permits are required if it is a major source, that is, it generally emits greater than 100 tons per year (tpy) of a regulated pollutant. For ozone nonattainment areas, lesser amounts classify a source as major. For example, in a severe ozone non-attainment area, a source releasing 25 tons per year (tpy) of VOCs or NO_2 is a major source. A gas turbine in a gas plant can emit over 100 tpy of NO_2 and be considered a major source. A gas plant processing sour gas will release SO_2 by flaring acid gas. If the plant is located in an attainment area for SO_2, then a permit is nevertheless required under the PSD program.

A review of the agency files should be undertaken to determine if any notices of violation (NOVs) have been issued, whether any fines have been imposed and if they have been paid, and what corrective actions have been required or what agreed orders have been entered. The buyer should be sure that all NOVs have been resolved and other required actions completed. If the acquisition is made, then the buyer must take the necessary steps with the appropriate agency to transfer the permit into his name.

A construction permit is required for new or modified sources. It is issued after a preconstruction new source review is completed under either the PSD program or the nonattainment area program. The preconstruction permit (new source review) for the prevention of significant deterioration or non-attainment programs imposes several requirements upon the applicant. Since the buyer assumes these responsibilities, he must be aware of the requirements. For the PSD program, these include, among others:

- The source must use Best Available Control Technology (BACT).

- Monitoring of ambient air quality must be performed for one year. (Hence, the buyer must ascertain if this is being done and, if so, how and by whom.)

A new source review is also required for a new or modified source in a nonattainment area. A facility may be in an area that is classified as attainment for one pollutant but non-attainment for another. If the source is in a nonattainment area, then some of the specialized requirements under the new source review are:

- Lowest Achievable Emission Rate (LAER) technology must be used.
- There can be no net increase in emissions. Necessary offsets must be provided, that is, the person constructing the source must reduce emissions from other sources so that there is at least a total offset plus some defined additional reduction. This provides a net decrease in emissions. The source of the offsets should appear in the preconstruction review permit.

Title V of the CAA Amendments of 1990 prescribes uses of an operating permit. At the time of this writing, the applicable regulations are in a process of formulation. The concept of the Title V permit is to incorporate into one "operating permit" all permit requirements. The requirement of the HAPs section will also be incorporated into the Title V permit. The Title V permit may be either a general or standard permit for a small source or an individualized operating permit for a major source. In all events, the need for a Title V operating permit must be anticipated.

Controls

The permits and the SIP must be reviewed to determine the permit control requirements for the facility. The facility should then be evaluated to ascertain first, that the required controls have been implemented, and second, that they are working. Control equipment is tested during startup to confirm that the emissions are within the permit. Particular attention should be given to older equipment, because it may no longer control the emissions to the permitted level. The maintenance records, performance tests, and records of emissions should also be reviewed. The basis for the emission quantities should be determined—were they actually measured? or were they calculated? or were they merely estimated? What was the methodology employed and what does the regulatory agency specify or recommend?

A physical walk-through should be made for visual observation of lines, valve, pumps, compressors, and vent stacks. Are the controls installed? Do they operate? Is non-emergency relief piping that does not go to another piece of equipment flanged at an "open" end?

Hazardous Air Pollutants

HAPs are defined in CAA Section 112, and a list of 189 substances have been classified as such by the act. The EPA Administrator can classify additional substances as HAPs. Many of these HAPs are found in normal exploration and production operations: benzene, toluene, ethylbenzene, xylene, n-hexane, DEA, ethylene glycol, formaldehyde, methanol, and polycyclic organic matter. A facility is a major source if it emits 10 tons per year of any one of the HAPs or 25 tons per year of any combination of the HAPs. A facility releasing lesser amounts is called an "area source." Major sources are subject to control technology requirements called Maximum Available Control Technology (MACT), while area sources are subject merely to Generally Available Technology (GACT). These technologies have not been fully defined as yet, but the prospective buyer must be aware of the potential requirement for retrofitting controls to the facility to comply with the Clean Air Act Amendments of 1990.

HAPs can occur from several locations in a facility: glycol regenerator vents, combustion sources, and fugitive emissions. Glycol dehydration systems are common in field operations and in gas plants. The regenerator heater generates NO_x and CO. For small systems, these may not be sufficient to result in major source status. The other principal emission is the vented vapor stream, which is predominantly water vapor but also contains benzene, toluene, ethylbenzene, and xylene (BTEX). BTEX is also absorbed from the natural gas stream by the glycol and is boiled off with the water. These compounds are regulated under the HAP and VOC provisions of the CAA.

The provisions of CAA Section 112, concerning HAPs, limit the applicability of this program to oil or gas wells. Note that Section 112(n)(4) provides that emissions for any oil or gas well (exploration or production well) and emission from any pipeline compressor station (or pump station) shall not be aggregated with emissions from similar units (regardless of whether they are in a contiguous area or under common control) to determine if they are a major HAP source. Further, for any oil or gas well (with its associated equipment) the emissions shall not be aggregated for any purpose under Section 112, nor are they to be listed as an area source by the EPA Administrator (unless production wells are located in a metropolitan statistical area with a population in excess of one million and the administrator determines that HAP emissions from these wells pose more than a negligible risk of adverse public health effects).

Section 112 Risk Management Program

HAPs are regulated not only because of normal emissions during operations, but also because of the potential for accidental release. An operator must determine if a Risk Management Program (RMP) is required under CAA Section 112(r). Although final regulations under Section 112(r) have not yet been adopted, the prudent buyer must anticipate their adoption. If the facility has an amount equal to or greater than the threshold amount set by the EPA Administrator, then the facility must have a risk management program. Propane and butane are listed with threshold amounts of 10,000 pounds. This corresponds to about 56 barrels of propane and 49 barrels of butane. If a facility has an amount greater than this, then it must have a RMP because of the flammability of these substances. The program requirements are discussed in more detail in Appendix 9. Compliance may require the use of outside consultants to conduct the required studies and prepare the required plans. A proposed rule has been published that would require: 1) a hazard assessment, 2) a prevention program, and 3) a response program. The hazard assessment alone encompasses a broad range of technical areas such as air dispersion and consequence analysis (such as, what is the concentration where people are located? what are the effects if an explosion occurs?). The prevention program imposes substantial requirements including standard operating procedures and change controls, and the use of risk assessment tools such as hazard operability analysis (HAZOP), failure mode and effect analysis, and fault trees. It is prudent to determine the applicability of the risk management program to the facilities being considered for purchase. The buyer should inquire as to the current status of the various components, including consequence analysis, risk assessments, and response plans. If these do not exist, a RMP must be developed after the EPA issues final regulations. Many of these requirements are similar to those of OSHA's Process Safety Management regulations.

There are proposed reporting requirements. The CAA requires reporting of releases of HAPs greater than a threshold amount for the previous five years. The prospective buyer should determine if the data is available for making such reports.

Nuisance

A nuisance is an unreasonable interference with another's use and enjoyment of his property. Emissions from a plant can pose such an interference. Examples of operations that could constitute a nuisance include: a gas odorant process, emissions of sulfur compounds, noise from plant equipment, and truck traffic (especially if the amount and type will change after acquisition). Generally, an established facility would not be a nuisance, but if the nature of the facility is changed by increasing volumes or if new processes are added generating new emissions, then it can become a nuisance. Further, even though the facility has been established and operating, the nature of the use of the surrounding property can change.

The surrounding area should be investigated to determine the nature of its use. If there are new, planned, or possible housing or commercial developments, there may be a potential for a nuisance to exist. A history of public complaints or opposition to the facility could indicate a high probability of such a problem. Facilities processing sour hydrocarbon fluids should be checked carefully for odor problems. Odor problems can also arise from processes that add odorant to the gas and from fugitive emissions.

Conclusion

The prospective buyer should determine if a permit is required or will be required. An evaluation of compliance with the permit is an appropriate next step. The current implementation of the operating permit will impact the facility. Another impact to be considered is the future implementation of the risk management program provisions of the amended CAA. By reviewing the requirements and the compliance status of a facility, a prospective buyer can determine possible additional costs to be considered in his decision.

California

It is sometimes supposed that if the Federal Clean Air Act regulations are satisfied, nothing more is required. However, while much of California's air quality regulation involves implementation of the Federal Clean Air Act, it is also true that businesses, including oil and gas operators, in California must also comply with California's own air pollution laws. The law is found under Health & Safety Code Division 26, Air Resources,[1] as amended by the California Clean Air Act of 1988 (CCCA).[2] The law is administered by the California Air Resources Board (ARB), which is part of the California Environmental Protection Agency, and by local or regional Air Pollution Control Districts and Air Quality Management Districts. Most of the day-to-day regulation of oil and gas operations occurs at the Air District level under Air District regulations, which vary from district to district.

In 1983, the California Legislature also enacted the Toxic Air Contaminates Chapter of the Health & Safety Code.[3] Here again, the actual control of toxic air contaminates has been delegated to the local air districts.

Air districts in the various oil and gas producing regions in California have adopted a variety of regulations that affect oil and gas producers. For instance, the South Coast Air Quality Management District, which is responsible for managing air quality in Los Angeles, Orange County and Riverside County, and the non-desert portion of San Bernardino County, prohibits anyone from installing, altering, or replacing any equipment, the use of which may either cause the issuance of air contaminants or may eliminate, reduce, or control those air contaminants without first having obtained a permit from the District.[4] It also prohibits such contaminants as certain visible emissions, creation of excessive fugitive dust, the use of certain fuels, etc.[5] Similarly, it issues Standards of Performance for various stationary sources of pollution, among them standards of performance for petroleum storage tanks.[6]

The lesson to be learned is that any oil and gas producer operating in California must identify the air district having jurisdiction over its operations and carefully review its rules and regulations. This is of concern not only to onshore operators but also to those who operate on the Outer Continental Shelf. This is so because federal law provides that air quality requirements applying to OCS sources located within 25 miles of the states' seaward boundaries must be updated periodically to remain consistent with the requirements of the corresponding onshore area.[7]

Louisiana

The Louisiana Department of Environmental Quality is developing a two-part permit scheme under Title V covering air emissions from oil and gas exploration and production facilities. Part one of the scheme involves a general permit to reduce the potential for emissions from sources that can be controlled below major source thresholds. Part two involves a general permit for sources subject to Title V permit requirements. The permit scheme could allow Title V permitting of Louisiana E&P facilities by the latter part of 1995.

Mid-Continent

The Oklahoma Clean Air Act and the Kansas Air Quality Control act have been enacted in part to implement the federal Clean Air Act. These state laws may affect oil and gas activities, such as hydrogen sulfide emissions from gas sweetening plants. The buyer should ensure that the properties are in compliance with these state acts.

CHAPTER 17 ENDNOTES

1. Health & Safety Code §§ 39000–44394.
2. 1988 Cal. Stat. Ch. 1568.
3. Health & Safety Code §§ 39650, et seq.
4. Rule 201, SCAQMD.
5. Regulation IV, SCAQMD.
6. Regulation IX, SCAQMD.
7. Clean Air Act § 328(a)(1), 42 USC § 7627(a)(1).

18

Representations and Warranties

Representations and warranties in contracts for the acquisition of oil and gas properties offer partial—but not complete—protection of the buyer from environmental liabilities. This is true also for contractual indemnities and covenants (see Chapters 19 and 20, respectively). It must be kept uppermost in mind that:

1. No private third party agreements will protect against enforcement action by the EPA (Environmental Protection Agency) or other governmental agency; specifically, no private third party agreements will prevent the EPA from seeking to hold all potentially responsible parties jointly and severally liable for CERCLA (Comprehensive Environmental Response, Compensation, and Liability Act) liability.
2. No private third party agreement can be used to defer or pass along criminal liability; the party found guilty in a criminal proceeding must shoulder his own burdens without regard to contractual representations and warranties, indemnities, or covenants.
3. Irrespective of private third party agreements, in many instances, permits and licenses issued by environmental agencies and authorities simply cannot be transferred to a third party; in other instances they cannot be transferred except upon satisfaction of conditions imposed by the issuing agency or authority.

There is nothing unique about oil and gas transactions that requires special treatment of environmental warranties and representations, indemnities, or covenants. Similar considerations apply as in any property transactions. Also, the same considerations apply whether the deal involves the purchase and sale of producing properties, the acquisition of an undeveloped prospect, the purchase of a single lease from a lessee, a farmout agreement, or the acquisition of a lease from a landowner. In each case, the relative bargaining strengths of the respective parties may vary considerably, and the acquiring party may not be in a position to insist on contractual provisions that will give it strong protection against environmental liabilities. This may be especially true in acquiring leases from landowners, many of whom will find environmental contractual provisions new, unfamiliar, and threatening.

This does not mean that environmental representations and warranties, indemnities, and covenants are any less important to the buyer or to the seller. There is a growing trend in oil and gas transactions that either party may be less willing to exercise negotiating flexibility on such contractual provisions and that this inability to agree on environmental contractual provisions can be a deal-breaker.

The buyer should seek environmental warranties and representations that the following are true:

1. The seller possesses all required environmental permits.
2. The seller has complied with all applicable federal, state, or local environmental laws; statutes; regulations; rules; ordinances; licenses; and permits relating to ownership, use, or operation of the property.
3. No pollutants, contaminants, toxic or hazardous wastes, or other hazardous materials have been generated, stored, treated, disposed of, discharged, dumped, located, or released onto or from the property or land contiguous to it.
4. The property is not subject to any investigation, or unsatisfied reporting or remediation requirements by any governmental body or agency.
5. The seller has no notice of violation or correspondence concerning real or potential violations from any governmental body or agency or any private party with regard to the environmental condition of the property.
6. The seller has no notice of any writs, injunctions, decrees, orders, judgments, lawsuits, claims, or legal proceedings pending or threatened relating to the environmental condition of the property.

7. No lien has been placed on the property pursuant to any environmental law, nor has the property been placed or is under investigation to be placed on the National Priorities List under CERCLA or any similar federal or state law.
8. All uses made of the property and all waste disposal practices on or from the property by the seller are as described in his written statement furnished to the buyer.
9. The property has not been designated as subject to special flood hazards or as a wetland.

The buyer can anticipate that the seller may seek limitations on warranties and representations such as these:

1. The seller makes no warranties or representations as to environmental matters, and the buyer agrees that he has performed environmental risk assessment to his own satisfaction.
2. The seller's warranties and representations are limited to "the best of seller's knowledge," "the actual knowledge of seller," or "the period of seller's ownership, control, and use of the property."
3. The seller disclaims all warranties and the property is conveyed "AS IS, WHERE IS," with additional language to the effect that the buyer agrees that no warranties or representations have been made by the seller as to environmental matters relating to the property and none has been relied upon by the buyer.

Although the buyer should resist the use of "as is" language, both he and the seller should recognize the severe limitations on its usefulness to the seller:

1. An "as is" provision is no protection against assertion of liability by the United States.
2. An "as is" provision does not protect the seller against liability to any subsequent owner beyond the immediate buyer.
3. An "as is" provision does not protect the seller against CERCLA statutory liability even as to the buyer.

4. Under CERCLA, a seller who has actual knowledge of a release or threatened release of a hazardous substance on the property (even though such releases occurred prior to the period of the seller's ownership, use, or operations) and who fails to disclose that fact to a buyer, will lose any right to defend against CERCLA liability on the basis of the "innocent landowner defense" (see Chapter 1).

California

It is sometimes thought that a seller who knows of a defect in his property may remain silent and sell the property to an unsuspecting buyer without risk of liability. However, this is not the rule in California. In California, if a seller has actual knowledge of a defect, knows that the buyer does not know of the defect and is not likely to learn of its presence in the course of a prepurchase investigation, the seller cannot remain silent. He must divulge what he knows or he is liable in a fraud action brought by the buyer for rescission or for damages.[1]

Louisiana

Precautions provided in this chapter regarding the use of "as is" language are particularly appropriate for transactions subject to the Louisiana law. Louisiana courts have regularly ignored the use of "as is" language, even to avoid liability being imposed on the immediate buyer, unless very specific supplemental warnings regarding defects in the subject matter of the transaction are contained in the transactional document.

CHAPTER 18 ENDNOTES

1. Lingsch v. Savage (1963) 213 Cal.App.2d 729.

19

Indemnities

CERCLA (Comprehensive Environmental Response, Compensation, and Liability Act) expressly encourages indemnity and allocation of liability agreements privately made between jointly responsible parties. This does not mean that an indemnity agreement will deter the EPA from asserting the joint and several CERCLA liability of the parties to the United States. Nor does it mean that the courts will take account of an indemnity agreement in finding joint and several liability in legal action by the United States, as distinguished from actions between the parties.

Several federal decisions, however, place specific conditions on enforceable indemnities in actions between the parties, which are difficult to rationalize but which nevertheless must be complied with. Although these particular decisions relate to indemnity for CERCLA liability, it is wise to comply with them in any environmental indemnity. Briefly stated, these requirements are that the indemnified should assign his liability obligation to the indemnifying party, which should assume the obligation and formally release the indemnified party.

Allocation of liability and indemnity may be made as of a specific date, such as the closing date. Indemnity can be limited to specific conditions, contaminants, or events, (such as indemnification only against remedial actions) or to a specified dollar amount. It can contain provisions concerning the respective rights of the parties to retain counsel and control litigation.

The parties may wish to limit their indemnity liability to a specified time period, which should be negotiated with an eye towards the statutes of limitations under environmental laws. Under the federal laws, these can range from three to six years and may not begin to run until remedial action is complete. In fact, the question of when remedial action is complete under EPA requirements can be vexing.

In any acquisition, it is appropriate for the seller to indemnify the buyer against any environmental violations occurring prior to the acquisition, and for the buyer to indemnify the seller against any environmental violations occurring subsequent to the acquisition.

In joint operations, it is appropriate for the party contributing a lease to indemnify the other parties to the acquisition against liability for environmental violations occurring prior to the acquisition. Whether or not the Operator indemnifies the Non-Operator against environmental violations occurring during the course of joint operations is a matter for negotiations between the parties. In any event, it is appropriate for the parties at least to agree between themselves that all environmental risks arising out of the operation will be borne in the proportions of their ownership of the joint operations property. This is a result of the joint and several liability of co-defendants in an action by the United States or by third parties, especially under CERCLA. Such agreement should be made enforceable by joint indemnity of each party as to the other.

In the special case of non-consent (full risk) operations, the participating party should bear the entire environmental risk arising in the course of the non-consent operation and should indemnify the non-participating party against the totality of such risk.

Louisiana

Louisiana's "mini-CERCLA" statute also preserves the enforceability of private indemnification and subrogation agreements.[1]

CHAPTER 19 ENDNOTES

1. La. R.S. 30:2276(I).

20

Covenants

Environmental covenants favorable to the buyer include the following:

1. Agreement to give the buyer the right to inspect the property and all documents, records, and permits in possession or control of the seller relating to: (i) environmental remediation, (ii) all uses of the property, presently and in the past, (iii) all hazardous substances placed on the property, and (iv) the conduct of waste disposal operations on or from the property.
2. Agreement that the buyer may discuss with any of the seller's employees or contractors, and with any governmental authority any environmental problems or potential problems disclosed or indicated by inspection of the property or by the seller's records.
3. Agreement that if the buyer's inspection or inquiry identifies any unremediated contamination of the property or other serious environmental problem, then the buyer, at his option, may: (a) terminate or rescind the agreement, (b) renegotiate the purchase price conditioned on remediation, or (c) require the seller at his expense prior to closing, to clean up the contamination and remediate any other environmental problem.
4. Agreement that the buyer may terminate the agreement if the seller is unable to obtain any environmental license, permit, or approval that the buyer deems reasonable and necessary to use the property in accordance with the intended purposes.

The buyer can expect that the seller will reasonably negotiate for covenants such as the following:

1. Limitation on the period of time during which the buyer may inspect the property and the seller's records, after which the buyer will be deemed to have waived questions not theretofore specified in written notice by buyer to seller.
2. Place a dollar cap not only on any cost for remediation that the buyer may require but also on any adjustments to purchase price or any amount to be escrowed.
3. Require prior consent to any discussion of problems by the buyer with any of the seller's employees or contractors or with any governmental authority, and require the seller's presence at any discussions.

21

Pre-Closing Investigation Of Liability Potential

The 1980s saw a frenzy of business acquisitions and mergers as companies sought to diversify their asset portfolios with the aid of Wall Street investment bankers, who were only too eager to develop new or expand existing financing mechanisms for such acquisitions. The more sober decade of the 1990s has seen a definite dampening of the buying spree, but even now many oil and gas companies are engaged in large-scale transactions in which multiple producing properties are bought, sold or traded in an effort to consolidate holdings and cut production costs.

The buying and selling of properties in these asset restructuring efforts have reached the point that many oil companies have established aggressive "acquisition and divestiture" teams to look for attractive business opportunities to redefine their property portfolios.[1] One major oil company has indicated that it will dispose of up to two-thirds of its domestic properties over the next five years.[2]

In years past, the period between the time the property was offered for sale and the closing of the transaction was used to draft the agreement and possibly to obtain financing. However, enactment of various environmental statutes has changed all that. Now, the time between offer and closing must be used not only to draft the agreement and obtain financing, but also to conduct reviews of the environmental conditions of the property. If a third-party lender is involved, the time is used to satisfy him as to any environmental problems present at the property for which the loan is being made.

The need to ascertain the environmental condition of a property prior to its transfer made obvious the need for environmental reviews. However, the environmental statutes are silent on the nature and extent of the environmental review needed to satisfy them or to alert the buyer sufficiently on the nature of any environmental liabilities being assumed.

Environmental Review Programs

Although there are many environmental requirements applicable to oil and gas properties, it was CERCLA's (Copmprehensive Environmental Response, Compensation, and Liability Act) strict, joint and several liability provisions that had the greatest force in impelling buyers and sellers alike to look at the environmental conditions of the property prior to its transfer. The focus of these early reviews was on avoiding CERCLA liability by laying the groundwork for the "innocent landowner defense."[3]

This defense requires the buyer to make "all appropriate inquiry" into the previous ownership and past uses of the property. Since the statute sets no standard for determining what must be included in such an inquiry, buyers are left to their own devices to design such investigations. These have come to be known as Phase I and Phase II assessments, depending on the nature and extent of the inquiry.

The Phase I assessment generally is limited to a visual inspection of the property; that is, no sampling of soils, water, or other environmental media is undertaken. Phase II assessments are aimed at confirming or ruling out the environmental concerns identified in the Phase I assessment. The nature and extent of the Phase II assessment depends, in large part, on the findings of the Phase I assessment, but it very likely will include some physical testing, such as air, water, or soil analysis. For example, if the Phase I assessment indicated large areas of stained soil, the Phase II assessment would likely involve soil sampling to determine the nature of the stain, the depth of its penetration, and (if groundwater were encountered) water sampling and analyses to determine whether and to what extent the water may be contaminated.

Phase I Assessments and the ASTM Guideline

Although CERCLA specifies that a buyer has to conduct "all appropriate inquiry" into the past uses of a property in order to establish the innocent purchaser defense, it does not provide any guidance on how that inquiry is to be conducted. In the absence of statutory guidance as to the type of the investigation needed, there grew up a host of investigative methodologies for conducting these Phase I assessments. Development of individual (sometimes property-specific or consultant-specific) assessment criteria resulted in a great deal of inconsistency, not to mention a lack of assurance that the inquiry would be deemed "appropriate" at a later date by a court being asked to shift CERCLA liability from the current to a prior owner.

In an effort to bring some much-needed order to the hodgepodge of formats used for Phase I assessments, the prestigious American Society for Testing and Materials (ASTM), founded in 1898 for the purpose of establishing test standards for materials, products, systems, and services, created the ASTM E.50.02 Subcommittee on Environmental Assessments for Commercial Real Estate.[4] The subcommittee has developed a standard practice for environmental site assessments as an aid to those seeking to establish an innocent landowner defense.

Scope and Purpose of the ASTM Guideline

The standard practice developed by the subcommittee consists of two standards: (i) the Transaction Screening Process and (ii) the Phase I Assessment standard. In theory, the prospective buyer will utilize the Transaction Screening Process first. This consists of obtaining answers to a host of questions about the current and past uses of the property to be acquired and the current and past uses of the adjoining property. In the ASTM's opinion, if the answer to any of the questions posed is "yes," the prospective buyer must then conduct a Phase I Assessment if he hopes to avail himself of the innocent landowner defense.

Now that the ASTM Guideline has been published, it behooves the environmental attorney to review his client's Phase I environmental assessments to determine how well they compare to the ASTM standard. Furthermore, when faced with a defendant's innocent landowner defense, counsel for plaintiff may well attack it by unfavorably comparing the defendant's environmental assessment with that described in the ASTM Guideline. It is appropriate to point out, however, that the ASTM Guideline was prepared specifically for commercial real estate transactions and not for oil and gas transactions.

Due Diligence Reviews

A Phase I environmental assessment is indispensable to a party seeking to ascertain the physical condition of a property at a moment in time and for laying the groundwork for CERCLA's innocent landowner defense. But a Phase I assessment cannot be expected to provide a complete picture of the environmental issues that will result from ownership of the property. For example, a Phase I assessment, by its nature, will not give the would-be buyer any information about the off-site waste disposal facilities utilized by the current owner and hence no inkling that the property may be the source of waste disposed of at a Superfund site. Ignorance of this fact can lead to Superfund liability for the new owner. In one particular case, for instance, a company purchased a tract of land years after wastes generated on that property had been sent out of state for disposal. The off-site disposal facility was listed on the National Priorities List of Superfund sites during the time that the property was owned by the prior owner; the prior owner was named as a potentially responsible party (PRP). This information was not disclosed to the new owner. After closing, the new owner was promptly named as a PRP on the basis that it was the owner of the property from which the waste had been generated. Arguments to the effect that the new owner had no connection with generation, transportation, or arrangement for disposal or disposal of the waste were to no avail with the Environmental Protection Agency (EPA), and the new owner eventually was forced to enter into an agreement to settle for a share of the cleanup costs.

To unearth the broader picture of the environmental liabilities connected with ownership of the property being acquired, rather than addressing merely the physical condition of the particular property, it is necessary to expand the environmental review from a Phase I-Phase II assessment into a more global "due diligence" or "business acquisition" review. For purposes of this chapter, the term "due diligence review" refers to the kind of review that looks not only at the physical conditions of the property (the Phase I and Phase II assessments), but also at statutory and regulatory compliance, litigation, and public relations history in determining the nature and extent of actual and potential exposure to environmental liability. The rest of this chapter is aimed at describing a "due diligence review" of an oil and gas property.

A Due Diligence Review of Oil and Gas Properties

Before a due diligence review is undertaken, it is important to ascertain the nature of the acquisition (assets only, purchase of stock, or corporate merger), since the nature of the transaction will affect the extent of the liability transferred to the buyer. If the transaction involves the merger of one company into another, the "surviving" company is well advised to conduct the most extensive review that time allows, since all the liability of the merged company will follow the assets. On the other hand, if the transaction involves the mere purchase of assets, the review may be more limited and may be directed only at ascertaining those liabilities for which the seller is providing no indemnification.

Having ascertained the depth of the review called for by the transaction, the buyer is in a position to determine the makeup of the due diligence review team. The more diverse the facilities being acquired, the larger the team (or the broader the expertise of each member) needs to be. In the context of an oil and gas property, for example, the assets being acquired may include producing wells, waste disposal wells, water wells, and monitoring wells, wells requiring plugging and abandonment, enhanced recovery facilities, undeveloped property, gas plants, compressor stations, cogeneration facilities, and interests in properties operated by others.[5] In an acquisition of this nature, the due diligence review team could well include environmental and safety (regulatory) professionals, scientists (geologists, toxicologists, biologists), engineering staff familiar with the process and equipment, reservoir engineers, landmen, tax specialists, oil and gas attorneys, environmental attorneys, and litigators.[6]

The role of these professionals in a due diligence environmental review will become clearer in the discussion that follows. For now, suffice it to say that each will review the appropriate portion of the facilities or records and assist in compiling the composite report to management.

The elements of a due diligence review discussed herein consist of checklists for: (a) the visual inspection, (b) records review, (c) litigation, enforcement, and public relations history, (d) interviews, and (e) report to management.[7] The questions and topics addressed in these checklists are not all inclusive. Whether or not using these checklists will constitute "all appropriate inquiry" for purposes of CERCLA and will be considered sufficiently "diligent" are questions requiring the advice of counsel in each specific case.

Visual Inspection

The due diligence review team should visit the site at least once, preferably while it is in operation.

The visual inspection should begin even *before the team reaches the fence line.* Things to look for or note include:

- Are smoke plumes visible from outside the fence line? Are odors emanating from the facility?
- Proximity to other facilities, whether industrial or residential. If industrial, categorize (e.g., manufacturing, food, exploration/production). If residential, note whether the homes are upwind or downwind in prevailing wind patterns. This is especially important if there is hydrogen sulfide in the hydrocarbons.
- Is the facility located in a poor, minority, or indigenous community? If so, be alert to potential citizen complaints based on allegations of environmental inequity, environmental injustice, or environmental racism.
- Access to the facility. Are the roads well maintained? Paved or oiled for dust suppression? Private or public? Are there warning signs? Condition of roadside ditches—do they look as if they have been used for disposal of industrial wastes, possibly from the facility being reviewed?
- General topography. Is the facility in a valley, on the plain, or on a hill? Where will stormwater flow—from the facility or toward it? Note type, quantity, and condition of natural vegetation close to but not at the facility. Note the same characteristics for the soil; note any evidence of natural erosion, particularly if vegetation is scarce.

Once inside the fence line, *tour all the buildings,* both offices and process-related. Even office buildings can contain or generate hazardous wastes.

- Is asbestos used in insulation?
- What is the source of the potable water? If bottled water is used, ask why. If the water is supplied by a private well, ask whether it is part of the property being purchased. Are other parties supplied from this well also?
- Look for comfort cooling towers. Are biocides used to treat the water to keep it from fouling the cooling tower used for

air conditioning? Many biocides are hazardous substances; their residues may be hazardous wastes.

- Are there emergency generators? How are they fueled? How often are they tested or fired? (i.e., how often are there air emissions from these generators and are the emissions accounted for in the plant's permit?)
- How are sanitary wastes managed? Is there a septic tank system on the property, or are wastes piped into a municipal sewer system?

Tour the process area, paying special attention to equipment, waste management facilities, and storage areas.

- Inspect process equipment and inventory it. At a gas plant, pay particular attention to the presence of such items as compressors, turbines, glycol dehydration units, and sulfur removal facilities. These may be subject to special requirements under such statutes as the Clean Air Act and its state and its local counterpart.
- Inspect the process equipment to ascertain the process used (for example, is the gas dehydrated using a lean oil process or is it dehydrated in a cryogenic process?) This will be informative in ascertaining the type and volume of wastes generated at the facility, as well as whether such wastes are hazardous.
- Inspect the process equipment to determine the level of maintenance exercised by the current owner. A clean, well-painted operation indicates care and attention paid on a regular basis. In contrast, equipment that is corroded and ill-maintained may indicate that the present owner has cut corners on maintenance, which in turn may mean there were more leaks, drips, spills, or other releases from poorly functioning equipment.
- Are there vents, flares, or other gaseous emission points such as pressure relief valves? These may trigger the need for permits or technological controls under the Clean Air Act.
- Look for asbestos insulation on process vessels. Is it exposed? Friable?

- Inspect the process areas for warning signs, such as "hearing protection required," "steel-toed footwear required," "benzene present," "H_2S (hydrogen sulfide) present." This will indicate how well the present owners know their operation, and whether they provided training and proper equipment to their personnel. This in turn may provide a clue to the plant's injury record, as well as pending litigation.
- Inspect wells and well pads. Look for signs of casing leaks, blowouts, fires, drips, or spills. Where and how were cuttings disposed of—on-site or off-site? Burial? Landspreading?
- Is there a pipe yard as part of the facility? Is there a special area for pipe containing naturally occurring radioactive materials (NORM)? Was the pipe simply stored in the yard, or was it cleaned and reconditioned also? (i.e., was scale removed from the pipe?) Look for evidence of scale on the soil surface. Also look for evidence that pipe dope dripped from the ends of pipe. Inquire as to whether and what solvents were used to remove pipe dope from the ends of pipe. Pipe dope may contain heavy metals that will test as "toxic" pursuant to the EPA's Toxicity Characteristic Leaching Procedure; many solvents are listed as hazardous wastes.

Inspect *storage facilities* such as tank batteries, drum pads, pits, warehouses, and the ubiquitous "equipment graveyard":

- In storage buildings, look for labelling on chemical bottles or other containers, Material Safety Data Sheets, signs on doors (radiation warning signs, halon warnings).
- At tank batteries, look for stains on the tank walls that may indicate overflows and spills. Are there standing pools of liquid in the area? Do the tanks have fixed or floating roofs? High-level alarms? Are the tanks diked and bermed appropriately?
- If there is an "equipment graveyard," look at the equipment that has been discarded or set aside. This will provide a clue as to past uses or processes, as well as possibly past accidents (e.g., a discarded vessel that is cracked or shows

signs of an explosion or fire). Look for discarded equipment that is still insulated—this may be asbestos. Look for electrical capacitors and transformers, which may hold PCB-containing oil.

- How are chemicals stored and handled? Are the drums labelled, or are there drums or other containers whose contents are unidentified? Are they in good condition? Are Material Safety Data Sheets available for all hazardous chemicals? Are drums sitting out in the open (exposed to the elements) or are they housed in a facility that has a good floor that provides containment for spills?
- Are there drip pans under spigots? Look for signs of spills or drips, such as stains on the soil or floor.
- Are waste drums and product drums held in the same area? Are hazardous waste drums clearly marked, segregated from others, and dated to indicate the length of time the waste has been stored?
- Where are empty drums stored? How are they managed? (reconditioned and reused? sold?)

Inspect all *waste management and waste storage areas,* whether part of the plant's waste management functions or "unofficial" dump sites utilized by the public.

- Inspect pits, ponds, lagoons, and other liquid waste management areas. Ascertain dates of construction, nature and volume of wastes managed therein, whether there are or have been any tests for leaks or leaching (i.e., groundwater monitoring studies), and the ultimate fate of the wastes managed in the pits (i.e., what happens to the supernatant? what happens to the sludge?). Are the pits earthen or lined?
- Inspect sites of pits that have been closed and backfilled. Inquire as to the nature of wastes that were managed in these closed pits. How long have pits been closed? How were they closed—dewatered, backfilled, and compressed? What happened to pit contents?
- Inspect disposal wells; inquire as to the nature and volume of wastes thus disposed as well as the injection pressures.

- Are effluents discharged into surface streams or impoundments, or are wastes discharged into a municipal or other wastewater treatment system? Is the discharge process water or storm water? Are there pre-treatment standards applicable to the discharge?
- Inspect solid waste management facilities. Where and how is office waste managed? Food waste? Has any asbestos insulation been disposed of on-site?
- Look for evidence of "midnight" or other unauthorized dumping. The materials may be no more hazardous than a threadbare sofa or an old mattress, but this could indicate that the public has viewed the property (especially the more remote locations) as an attractive place to dump their discarded materials. This in turn could also invite dumping of industrial or hazardous waste.

When inspecting the facility, look at the *soil surface, vegetation, surface creeks, or other natural impoundments.*

- Are there stains of any kind on the soil? Pay special attention to the soil surface near process equipment. Look for oily or dark stains. Ascertain whether they are the result of intentional application of oil as a dust suppressant, or of spills or leaks. Determine the nature of the oil used (e.g., tank bottoms, used crankcase oil, other).
- Are there salt deposits or crusts on the surface? This could indicate spills of produced water, which would be high in NORM as well as chlorides at certain times.
- Around a cooling tower, look for yellow stains that may indicate leaks of chromate-containing cooling water.
- Are there drains, sumps, or other liquid collection points? What effluents or discharges flow or drain into these?
- If gravel, shell shards, or some other material is used as a surface cover, note whether it has been freshly applied or if it looks weathered. A "too clean" appearance may indicate that the surface cover was added recently to hide stains or other signs of contamination.
- Look at the vegetation or lack of it. If vegetation is lacking, inquire whether herbicides are used for weed control or

whether or not the soil will support vegetation. Compare vegetation types and quantities inside the fence line with those noted prior to entering facility. Look for "bald spots" in the vegetation pattern, especially if they coincide with the site of closed pits, equipment that has been removed, spills of produced water, or places where sludges and scales were disposed.

- Note the appearance of any water in surface streams and impoundments. Note the condition of the water upstream and downstream of these discharge points. Are the banks clean or do they indicate that the water contains oil and grease or other contaminants?
- Is the land or are parts of it marshy or boggy, indicating a need for special investigation to determine whether it is wetlands?
- Are there any generally known or believed to be any endangered plant or animal species on the property?

Records Review

The purpose of the records review is: (a) to obtain documentation to substantiate or disprove what may have been a first impression formed during the visual inspection; (b) to establish a compliance history of the plant insofar as the records are available and accurate; and (c) to establish whether the appropriate records have been kept.

The list of records to review will vary from facility to facility, from one type of property to another, and from state to state; however, the following is a list of records that relate to the most common environmental requirements as they apply to oil and gas properties. State requirements will dictate the need to review other records mandated by state law and regulations. As with the checklist for the visual inspection, this list is not exhaustive, and the buyer should consult with counsel to determine whether additional records should be sought and reviewed.

Examine permits and authorizations:

- Clean Air Act (or state equivalent) permits to construct and to operate equipment that emits air pollutants.

- Clean Water Act (or state equivalent) permits for discharges of effluents, including both process water and storm water. Discharge monitoring reports.
- Generator identification numbers for hazardous waste generators.
- Hazardous waste disposal facility (TSDF) permits.
- Corps of Engineers "dredge and fill" permits pursuant to Section 404 of the Clean Water Act.
- Safe Drinking Water Act Underground Injection Control (or state equivalent) permits for disposal well(s).
- Waivers or exemptions under any of the applicable federal environmental statutes and regulations.
- Permits, waivers, or exemptions required pursuant to state law.
- Regulatory approvals to construct equipment or conduct operations subject to technology-driven non-permit programs such as those under the Clean Air Act.
- Registrations for pressure vessels, underground storage tanks, or other equipment.
- Review permits to ascertain expiration dates, dates for payment of permit fees, and requirements to transfer permits from one owner to another.
- If permits are not located, inquire as to the reason.

Examine applications for permits and other authorizations and permit compliance documentation:

- Applications for the types of environmental permits and authorizations noted above; determine whether any permits were denied; review the parameters specified in the permit applications, since these become part of the permit itself although not explicitly reflected in it.
- Notices to regulatory agencies certifying compliance or reporting malfunctions, upsets, by-passes, permit deviations, or other instances of self-reported noncompliance.

- Compliance testing reports, such as source testing for air emissions and discharge monitoring reports for water quality permits.

Examine other records:

- Hazardous waste manifests, solid waste manifests, or other required documentation.
- Emergency and contingency plans, such as Spill Prevention Control and Countermeasure (SPCC) Plans, SARA (Superfund Ammendments and Reauthorization Act) Community Right to Know emergency response plans.
- Documentation relating to required training for employees; attendance lists for training courses, safety meetings, etc.
- Material Safety Data Sheets for chemicals used by the facility or brought to it by contractors.
- Documents related to OSHA's (Occupational Safety and Health Act) Process Safety Management rule, such as process hazard analyses.
- Reports filed with the Securities and Exchange Commission describing environmental liabilities.

Litigation, Enforcement, and Public Relations History of the Facility

- Were environmental audits or assessments conducted by or on behalf of the current owner? Are copies of those reports available? If problems or violations were detected, were they promptly corrected?
- Are there any notices of violation from state, local, or federal regulatory agencies?
- Are there any judicial complaints, administrative orders, injunctions, unilateral or consent decrees pertaining to non-compliance in safety, health, or environmental matters? Have all suits been settled and orders complied with or are any of these still outstanding?
- Are there citizen suits (where a plaintiff is acting on behalf of the attorney general to compel compliance) or private

party suits for damages or threats of such citizen suits or private party suits alleging non-compliance with environmental requirements?

- Are there citizen complaints pending before any regulatory agency in connection with alleged discharges, emissions, contamination, releases to the environment, or nuisance conditions?
- Has enforcement action been taken on the basis of a citizen complaint?
- Has the seller been alleged to be a PRP pursuant to an action under CERCLA,[8] or alleged to be a responsible party pursuant to an action under the Oil Pollution Act?[9]
- Are there toxic tort or damage suits pending in connection with the events that led up to the seller's designation as a PRP, such as toxic tort litigation filed by homeowners living around a Superfund site where the seller is a PRP?
- Are there judicial complaints or unilateral or consent decrees pertaining to toxic tort suits in connection with emissions, discharges, or releases from the facility?
- Are there toxic tort suits still pending? Are there suits for damages to property or natural resources (e.g., groundwater) pending?
- Are there environmental liens to secure statutory obligations to any governmental agency filed against the property?

Interviews

The review team should interview the plant or facility manager and his employees, as well as management and environmental and safety professionals of the company that owns or operates the facility. The objective of interviewing plant personnel is to supplement, expand, or obtain clarification of any information gleaned from the visual inspection and the records review, as well as to get a historical perspective on past uses of the facility.

Some of the questions to pose to management and staff of the owner and/or operator of the facility include the following:

- History of the plant. When was the plant constructed? When were modifications, revisions, or expansions made?

Why? When and why were process changes made? Did emissions increase as a result of any of the changes?

- How was the plant operated? Are flow diagrams or plant schematics available? Are there aerial or other photographs of the plant and surrounding area?
- Inquire about the organization of the management, staff, and the safety and environmental professionals. If the seller did not have environmental health and safety professionals on staff, has it engaged qualified outside consultants?
- If the seller had safety and environmental professionals, were they dedicated solely to the facility in question, or did they have responsibility for a number of other facilities as well? Inquire about the role of these professionals in reviewing such things as process changes and equipment revisions.
- What was the mechanism for identifying new laws and regulations applicable to the facility?
- Ask about the owner's philosophy and corporate policy on environmental compliance. Does it seem to comport with the buyer's? If not, it may mean that some requirements were not addressed at the level the buyer will require.
- Is the seller a large corporation with a good environmental track record and a good reputation with the regulatory agencies? Or is the seller a sole proprietor or a small independent or some other owner not generally conversant with environmental requirements?
- Ask about the company's engineering and design standards, with particular emphasis on those areas not governed by specific codes or regulations. Does the seller follow trade association guidelines or customary industry practices where there are no mandatory codes?
- What mechanisms were employed to detect, report, and remediate instances of non-compliance with safety, health, and environmental laws, rules, and regulations?
- Inquire about any missing or incomplete records, permits, authorizations, reports, or other data.
- Inquire as to whether there are any complaints, problems, or other conditions that the interviewee knows about but

that are not reflected in the written records or apparent during the visual inspection.[10]

Neighbors of the facility must be interviewed also, whether they are residents or commercial tenants or industrial tenants. The objective of these interviews is twofold: (1) to learn whether they have any complaints about the operation of the facility (these complaints may indicate threatened litigation); and (2) to learn about past uses of the property. Some questions to ask neighbors include:

- Have they been bothered by odors, fumes, emissions, or discharges from the plant? If so, what have they done about it? Have they complained to any of the regulatory agencies? Have they complained to the company? Have they received any responses from either of these entities?
- Have they ever filed suit against the company for any of these problems? Do they know of anyone who has?
- Do they or does anyone in their family work at the facility or another facility owned by the same company?
- If the neighbors are long-term residents of the area, ask about the history of the plant and how this has affected the area. This will be indicative of whether the plant is viewed as a positive or negative influence in the community.

Records and Personnel Interviews with Regulatory Agencies

It is desirable to review records and interview personnel of the regulatory agencies with jurisdiction over the facility. The objective of these activities is to ascertain the compliance record of the facility and to find out, if possible, whether any of the problems uncovered are serious enough to be considered a legal violation, which will then entail an enforcement action. This element of a due diligence review may easier be said than done. Records of some regulatory agencies may not be available unless an advance request has been made and the agency has reviewed it, has made any secrecy or privilege determinations necessary, and has isolated or copied the relevant records. This process can take weeks or months. Interviewing agency personnel may be even more difficult.[11]

Report of Assessment

The final step in the assessment process is writing and presenting the report to management. As with any other report memorializing an investigation, there a number of standard elements must be included:

- Identify the members of the team participating in the investigation—their credentials, background, role in the investigation, and a means of communicating with them. Consider asking the team to sign the report. This will help determine whether there was consensus among the group as to the conclusions in the report and as an invitation to prepare a "minority report" in case there is a difference of opinion.
- Describe the property investigated by street address, legal description, or by name of the plant. A generic description, such as "the property being purchased from Company X," will prove insufficient at a later date when it may be necessary to establish that a specific piece of property was indeed investigated before being acquired.
- Describe the dates of the investigation. This will help to establish, among other things, the thoroughness of the inquiry.
- Describe the visual inspection in summary form. This will establish whether any sampling was done, whether the facility was in operation at the time or had been deactivated, as well as other conditions that may have hindered full investigation, such as weather conditions.
- Identify the people who were interviewed, listing their address, how they can be contacted, and their status (e.g., neighbor, government official, plant personnel).
- Describe the categories or types of records reviewed, and where they are located.

The heart of the report undoubtedly, is the portion dealing with facts, findings, and conclusions. Prior to writing the report, the members of the investigative team will find it useful to discuss proposed findings in order to ascertain the validity of their individual impressions. For example, the impression of the team member who reviewed records and found that the files were devoid of any environmental permits (which could indicate a serious problem if permits were required but had not been obtained), will be tempered by the information from a team member who participated in the tour of the facility, where he observed copies of the permits posted in the control room so they could be reviewed by the operators.

Care should be taken to keep the report as factual as possible, indicating if a statement is a firsthand observation (e.g., "the investigative team observed . . .") or is not (e.g., "the investigative team was told by Neighbor X that . . .").

In drafting conclusions, it will be advisable to obtain the assistance of counsel if the report will allege that there are violations of any legal requirements. In this situation, care must also be exercised to ensure that the report to management is informative and factual without reaching unwarranted conclusions. For example, it may be more accurate to state "the investigative team did not find, nor could plant personnel provide the team with copies of hazardous waste manifests, although the team was told that such documents are kept for three years," rather than saying "this facility did not manifest its hazardous waste."

The investigative team will want to consider making a recommendation as to the need for a Phase II assessment. A recommendation for such an assessment should indicate the areas to be investigated in greater detail, the nature of the problem needing verification, and the type of investigation being recommended (e.g., soil sampling, groundwater testing, emissions testing).

Finally, the report should include copies of any photographs, videos taken, or other supporting documentation.

Conclusion

Due diligence reviews provide buyers of oil and gas properties with a better picture of the environmental liabilities that may attach to the ownership of property. A conventional Phase I environmental assessment, while invaluable in ascertaining the physical condition of a property and in laying the predicate for a CERCLA innocent landowner defense, provides insufficient information to make a truly informed decision as to the nature of the liabilities that can be expected after the property is acquired. The due diligence review is also a much better tool for uncovering information that will assist the buyer in negotiating with the seller on the terms and conditions of the purchase and sale agreement in regard to environmental liabilities.

CHAPTER 21 ENDNOTES

1. Forest Oil Announces Results, Bus. Wire, May 11, 1992.
2. Chevron Schedules Big Changes, Oil & Gas J., Jan. 20, 1992, at 29.
3. 42 USC §§ 9601(35)(A), 9601(B).
4. This is a Subcommittee of the ASTM Committee on Environmental Assessments. Personal Communication with ASTM, December, 1992.
5. It is beyond the scope of this chapter to discuss the nature of environmental liabilities that may attach to ownership of various types of royalty or mineral interests. See Chapters 1 and 4.
6. For a truly comprehensive audit, the team may include other disciplines, such as financial auditors, royalty accountants, medical personnel, labor and employee benefits attorneys, patent counsel, etc. See John M. DeMeester, Practical Guidance for Due Diligence Environmental Auditing, 18 Envtl L. Rep. 12,210 (1988).
7. Each of the items in the checklist should be addressed in both a historical context (i.e., how were things done in the past?) and in the context of current operations.
8. 42 USC § 9607(a).
9. 33 USC §§ 2701–2761.
10. With the advent of criminal sanctions in the environmental arena, it may be difficult to elicit this type of information, since the interviewee may be concerned about personal liability.
11. A regulatory agency will not discuss any environmental action that is only under consideration or where disclosure of any information will be subject to certain strictures, such as the Privacy Act (5 USC § 552a) and the Freedom of Information Act (5 USC § 552).

22

Performing Cleanup Operations

Cleaning up a contaminated site requires the services of a knowledgeable environmental professional to assess the extent of the contamination in area, the depth to which the soil has been contaminated, and the potential involvement of groundwater, as well as the nature of the contaminating substances. Remediation or removal techniques are varied, both in effectiveness and cost. Although the technology is constantly improving, as a consequence, many technologies remain in an experimental stage. Perhaps more importantly, the contamination at any particular site is unique to that site, and cleanup requires remedies fashioned by the environmental professional to the site's particular problems. Generalizations as to cleanup operations consequently cannot be made.

It may be helpful, nevertheless, for the buyer to have at least some idea of the time and monetary risks involved in deal-making options. The options a buyer may consider to place the cost and risk of contamination cleanup or remediation on the seller include:

- Contractual requirements for cleanup prior to closing
- Seller's affirmative covenant to perform cleanup after closing, perhaps with the escrow of a portion of the purchase price conditioned on cleanup
- Seller's indemnity for cleanup costs
- Buyer's assumption of all or an acceptable portion of the risks and costs of cleanup, if necessary to prevent a deal from coming apart

This chapter is intended only to point out remediation alternatives that may be considered if contamination has been established.[1]

The options for remedial action fall into three general categories, which will be influenced by the type, extent, and degree of the contamination, as well as by the available time, space, and money. These categories for remedial action are:

1. To do nothing and let nature treat the contaminant over time. Professionals call this "passive remediation."
2. To treat the contaminant on the property, either in place without removal from the ground (called "in situ remediation") or by removal and treatment in a prepared area on the property and, if permitted, return to the excavation after treatment.
3. To remove the contaminated soil or material from the property and dispose of it in a proper and permitted manner (called "off-site remediation").

Passive Remediation

It is quite possible, and not infrequently the case, that the contamination is so minimal that it does not exceed regulatory thresholds mandating cleanup and may be considered to be immaterial for purposes of a sale. Most state and federal regulatory agencies have adopted risk assessment procedures setting the thresholds of contaminant levels allowable in the soil or ground water. Additionally, in cases of detecting O&G (oil and grease) or TPH (total petroleum hydrocarbons), the regulatory authorities may be flexible on a case-by-case basis in determining allowed levels for these contaminants, depending on information and tests for their specific components. Particularly in multiple-property acquisitions, specific properties with only a minimum of contamination not requiring remediation should be identified.

On-Site Remediation

On-site remediation alternatives involve the removal or reduction of contaminant concentrations in the soil through various physical, chemical, or biological means that do not require removal of the affected soils from the property. On-site remediation may involve either: (a) removing the soil from the ground for treatment in an area near the excavation and, where permitted, returning the treated soil to the excavation; or (b) treating the affected soils in place without removing the soil.

The decision on which technology is used will depend on the space available for surface treatment, if any, the time required to complete the treatment, and the type and extent of contamination involved. Also, the interests and attitudes of adjacent landowners may influence the decision to use surface treatment. In general, surface treatment may provide quicker results than treating the soil in place.

The on-site surface remediation alternatives include landfarming, bioremediation, and on-site thermal desorption treatment. On-site technologies are especially important in site conditions that preclude the physical removal of the contaminated soil, such as developed areas with physical structures that cannot be removed or sites hard to access with earth-moving equipment. There is a risk in any on-site technique that the rate of decrease in contaminant levels resulting from the particular technology selected may become less effective over the period of treatment. Thus, it may not be successful in attaining the required regulatory cleanup levels.

Landfarming involves the removal and controlled application of contaminated soils to a designated area for treatment whether on the property from which the soil has been removed or at an off-site location. The treatment may include bioremediation and volatilization of hydrocarbons compounds and attenuation of metals within the soil matrix. The technique has proven effective through many years of use in the petroleum industry, particularly for treatment of most petroleum hydrocarbon compounds. It does not require elaborate equipment or large capital costs for implementation. It does require, however, relatively large treatment areas. Some regulatory agencies require synthetic or clay liners beneath the treatment area as well as routine monitoring and reporting. There is a growing trend among oil and gas operators to join in the permitting and use of a common landfarming site for treatment and disposal of contaminated soils from operations within the same field or area.

Bioremediation of soil involves the degradation of hydrocarbons in the vadose zone of the soil (depths above the permanent groundwater level) by indigenous microorganisms. The naturally occurring biodegradation process is enhanced through the addition of oxygen and nutrients such as nitrogen and phosphorus. It is similar in principle to bioremediation of marine spills, but in both cases the technology is still being developed and refined. The technique can remove heavier hydrocarbon compounds, such as heavy crude, as well as lighter petroleum hydrocarbons. Soil bioremediation, however, is typically a long-term process.

Thermal desorption treatment involves the removal of hydrocarbon compounds from contaminated soils by heat-enhanced volatilzation. Excavated soils are processed through and heated in a closed chamber. The off-gases are then collected and routed through an air emissions or hydrocarbon recovery system.

Soil vapor extraction involves the removal of volatile hydrocarbon compounds from the vadose zone through physically drawing or venting air through the soil matrix. The basic components of the system typically include extraction wells, high-vacuum pumps, treatment units that control air emissions, and various monitoring systems. Soil vapor extraction is a proven technology. It has wide applicability, not limited by the size of the area or by the volume and type of soil affected. It is conducive to use in treatment of contaminated soils around or beneath buildings or structures. On the other hand, the technique removes only volatile hydrocarbons, such as gasoline and possibly diesel. It is not effective in low-permeability, heavy textured, or stratified soils, such as clays. It is typically a long-term process.

Soil washing and leaching involves injecting or forcing water downward through the vadose zone to remove hydrocarbons and drive them into the ground water, which is then remediated by pump and treatment techniques. Although this technique removes a broad range of hydrocarbon and other compounds, if the ground water is presently free of contamination, forcing hydrocarbons into it may create problems where none previously existed and usually require extensive permitting approvals plus routine monitoring and reporting.

Vitrification involves the use of electric currents to generate extremely high temperatures for the conversion of contaminated soils into chemically inert and stable glass or crystalline material. Unlike most technologies, vitrification is effective for the treatment of metal-contaminated soils and is especially applicable to crude and tank bottoms. The treatment can be accomplished in a shorter time frame than soil vapor extraction or soil bioremediation, but it is highly experimental, and expensive. The depth of treatment is limited to approximately 50 feet. The efficiency of the technique decreases with increase in moisture content of the soil.

Off-Site Technologies

Off-site remediation of contamination by the various technologies for treatment involve the excavation and removal of the affected soils for disposal or treatment and further disposal away from the original site. For smaller quantities of affected soil or for occasions where time is of great importance, off-site disposal is often a more cost effective and time efficient alternative.

Enhanced bioremediation uses the same treatment processes as traditional land treatment but in a more carefully engineered and managed manner. It is a proven technology and is effective for treatment of most petroleum hydrocarbon compounds. This technique involves moving the affected soils into specially prepared treatment areas or cells to undergo specifically designed enhanced treatment through application of selected microorganisms responsive to the particular contamination, selected nutrients and controlled oxygen demand requirements. It requires less surface area and is typically faster than traditional land treatment. On the other hand, it requires greater construction and operating costs than traditional land treatment. For these reasons, when the area of soils impacted by petroleum hydrocarbons is relatively small, enhanced bioremediation treatment usually is only possible at off-site treatment facilities serving a permitted class of industry. Such facilities treating petroleum affected soils are often operated by landfill waste management companies on properties adjacent or near to municipal landfills. After treatment, the soils treated to remove petroleum product are moved to the landfill for use as daily cover.

Incineration involves the destruction of hydrocarbon compounds through heating to extremely high temperatures in a closed chamber. The destruction and removal efficiency for hydrocarbon compounds in the soil usually exceeds 99.99%. This is a proven technology with high production rates, and greater predictability and treatment efficiency than many other techniques. Mobile processing units are available. It is, however, an expensive process. Administrative permitting is usually required, and there may be public opposition.

Asphalt incorporation involves the incorporation of petroleum-contaminated soils into hot asphalt mixes as partial substitute for aggregate. This technique is effective for treatment of most petroleum hydrocarbons and is unique in providing beneficial reuse of contaminated soils for road base or similar applications.

Solidification and stabilization involve the immobilization of waste through the addition of cement and other materials prior to ultimate disposal. This proven technology has been applied commercially at large-scale remediation projects at refineries and petrochemical plants. Landfill disposal involves transportation and disposal of the soil into a landfill permitted to accept soils contaminated to certain levels of petroleum hydrocarbons. Generally, where the soils are classified as non-hazardous and laboratory tests for TPH detect less than 1500 parts/per million, the soil may be accepted into certain municipal landfills for use as "daily cover" for the minimum tipping fee ranging from $5 to $15 per cubic yard. For higher levels of TPH, specially permitted landfills are available for more expensive tipping fees. Also, the costs of transporting soils to these specially permitted facilities may be expensive, depending on the regulatory transportation requirements.

CHAPTER 22 ENDNOTES

1. For a more detailed description of remediation technologies see monograph, Bioremediation of Soils Contaminated by Oil and Gas Operations, published by Kevin Van Hook, K. W. Brown Environmental Services, Houston, Texas, in 1993 and authorities therein referred to.

23

Corporate Insulation against Liability Exposure

In order to reduce *operations risks* in oil and gas exploration and production that may result from operating simply as a working interest owner, it is common to use a separate operating entity. Such entities include operating through: (a) a separate wholly owned corporation; (b) (to avoid adverse income tax consequences) a sub-Chapter S corporation; (c) a limited partnership; or (d) a limited liability company (which has been only recently conceived as a form for doing business and is untested in judicial decisions). In joint operations where the parties are acting as working interest owners, reduction of operating risks is sought by contractual controls in the joint operating agreement. Such controls include a provision that each party will be liable only to the extent of its proportionate ownership interest in the jointly owned property and a provision that the operator itself will not be liable to non-operators except for gross negligence or breach of the express provisions of the joint operating agreement.

Management of *environmental risks,* however, often cannot be achieved successfully by the same techniques. In joint operations, for example, CERCLA (Comprehensive Environmental Response, Compensation, and Liability Act) joint and several liability cannot be avoided by the non-operator's contractually casting the full responsibility on the operator, even though the operator may be the acting party whose conduct has resulted in the CERCLA violation. Furthermore, a joint operating agreement provision for sharing liabilities proportionately does not prevent the assertion of CERCLA liability against any one or more of the parties for the entirety of the cost of the cleanup. For more information, see Chapter 3.

The technique of operating through a wholly owned corporation is based on the common law concept that a corporate shareholder (including a parent corporation) is not personally liable for the torts or other liabilities of the corporation. This principle has been restricted, however, by the doctrine that where the operating corporation is merely the alter ego of the shareholder or parent corporation (as, for instance, where the operating corporation is undercapitalized or where the management and record keeping of the operating corporation have been confusingly intermingled with the management and record keeping of the shareholder or parent), a court will pierce the corporate veil and hold the shareholder or parent liable for the operating corporation's torts.

There have been a few federal court decisions in which these common law rules were also applied to liability for the corporation's violation of environmental laws. The majority of federal court decisions, however—and the clear trend of current decisions—is that the shareholder may be personally liable for the corporation's environmental violations. In some cases, these decisions are based on the fact that the shareholder actively participated in the environmental violation and therefore was liable under CERCLA as a statutory operator (as distinguished from owner), or that the shareholder had the corporate power to prevent the violation. In other cases, the court has simply avoided discussion of the common law doctrine of freedom of the shareholder from liability for the corporation's torts. The current trend of the cases is to hold that this common law doctrine does not apply to environmental violations, as a matter of an overarching public policy in favor of strictly enforcing environmental regulation.

It follows that, even though adverse income tax consequences are disregarded, use of a corporate operating vehicle is not a satisfactory way to manage environmental risks. Since a sub-Chapter S corporation is basically a corporation that has been modified under strict regulation to avoid adverse income tax consequences, the same conclusion applies to it

A limited partnership possibly insulates the limited partners against environmental violations by the partnership. However, there is no judicial authority on the question. According to the entity theory of modern partnership statutes under CERCLA, the partnership and not the partners are the owners. In order to preserve the limited liability of the limited partners generally, they are proscribed from entering into management decisions, and therefore could not be held to be statutory operators. The general partner, of course, is personally exposed to environmental liability. If, under IRS (Internal Revenue Service) regulations governing the required capitalization and other requirements for favorable tax treatment, the general partner itself is a corporation, its shareholders nevertheless may be personally liable for environmental violations for the reasons previously discussed.

During the past few years, limited liability companies have been authorized under the statutes of most states, and the use of this form of business entity in oil and gas exploration and production ventures is becoming increasingly popular. Their viability in protecting all members and managers from personal liability for the general obligations of the company remains untested by judicial decision, which is also true for environmental liability. Under CERCLA, almost certainly all statutory managers of a limited liability company may be held personally liable, notwithstanding the provisions of a state statute to the contrary, if they actively participate in management decisions involving environmental matters or if they have the corporate power to participate. Furthermore, as previously discussed, since, with respect to environmental violations, the federal courts have had little trouble ignoring the common law doctrine that corporate shareholders are not personally liable for acts of the corporation, it is predictable that the same courts will have little trouble ignoring statutory provisions that managers and members of a limited liability company are not personally liable.

24

Insurance

It is a common occurence that policyholders facing environmental liabilities are not aware of the insurance coverage that is potentially available in their existing insurance programs. This lack of awareness is particularly true for small businesses that do not employ risk managers or other insurance experts. When discovering a polluting event, the policyholder may, at best, contact his insurance agent and report the problem. Too often, the agent simply repeats the standard insurance industry reply that pollution is not covered, and the policyholder does nothing more.

Are environmental liabilities covered by insurance? Despite what insurers routinely tell their policyholders, the answer is "yes." The law, at least in most states, does not foreclose coverage. Therefore, most policyholders reasonably can expect coverage for most environmental liabilities.

At the onset, it should be understood that since environmental law is rooted in federal statutes as construed by federal courts, the significant features of environmental law are more or less uniform in all states. Legal issues involving insurance protection, including protection against risks of environmental liabilities, however, are uniquely grounded in the laws of the various states, and therefore vary from state to state. There is no federal law of insurance, either statutory or common. Thus, the law governing a particular case will be the law of a particular state. As will become evident from the discussion in this chapter, the judicial holdings also vary widely among the states.

Comprehensive General Liability Policies

The most common type of insurance that can provide pollution coverage is the conventional comprehensive general liability (CGL) policy. The standard-form CGL policy insures against liability to others on account of *property damage* or *bodily injury* caused by an *occurrence* except as expressly excluded by the policy. It also requires the insurer to defend the insured in litigation. Almost every business maintains commercial CGL coverage providing indemnity for, and defense against, property damage and bodily injury claims by third parties that arise out of routine business operations.

CGL policies exclude a number of risks, including those related to: (a) the policyholder's own property, (b) liabilities to employees, and (c) automobiles. These risks are excluded because separate lines of coverage generally are available in the marketplace (e.g., property insurance, workers' compensation insurance, automobile insurance). Unless specifically excluded, however, all other risks are generally covered by the CGL policy.

Requirement of Occurrence. Beginning in 1966, the standard-form CGL policy provided coverage on the basis of an occurrence rather than, as in previous policies, an accident. An occurrence was defined in the 1966 CGL policy as:

> *an accident, including injurious exposure to conditions, which results, during the policy period, in bodily injury or property damage neither expected nor intended by the insured.*

This definition broadened the scope of coverage by recognizing injuries resulting from *long-term exposure to conditions* (as distinguished from an abrupt event)[1] and by making it clear that the policy covered *the unintended consequences of intentional acts.* Both concepts are important in application of the coverage to environmental risks. Thus, the coverage extends to damage caused by gradual pollution events, as long as the resulting injuries or damages were neither "expected nor intended" by the policyholder, notwithstanding that the release or discharge of pollutants may have been intentional. This distinction is important since much of today's pollution resulted from lawful and intentional discharges of pollutants prior to the time that society was aware of the potential consequences of those acts.

The issue of whether such damages were "expected" by the policyholder is troublesome. Most courts refuse coverage when a policyholder in fact subjectively expected the pollution damages to occur. For example, in a California case, the court held that the appropriate test for "expected" damages is whether the insured actually knew or believed its conduct was substantially certain or highly likely to result in damage.[2]

At the other extreme, some courts apply a purely objective test to determine if the policyholder knew, or should have known, there was a "substantial probability" that certain results would follow. In Meridian Oil Production, Inc. v. Hartford Acc. & Ind. Co., for example, the Fifth Circuit U.S. Court of Appeals held that under Texas law, pollution damages arising from Meridian's oil and gas activities did not constitute covered occurrences because Meridian's "conduct inevitably and predictably caused the pollution."[3] The court evaluated Meridian's conduct based not on its subjective knowledge, but on an objective "reasonable prudent operator" basis. This approach effectively eliminates coverage for a policyholder's negligence, the very thing for which insurance is purchased.

In cases involving unexpected and unintended pollution damages resulting from "negligent" or unintentional acts, coverage should be upheld. In cases involving unexpected and unintended pollution damages resulting from intentional waste discharges or disposal practices, the outcome is less clear.

Regardless of the standards applied, the more involved the policyholder is with the actual pollution, the more likely it is that courts will deny coverage on the basis that the pollution was "expected or intended." Therefore, a policyholder who is the owner or operator of a facility that discharges hazardous substances into the environment (an "active" polluter) is less likely to sustain coverage than a waste generator who sends waste off-site to a facility where pollution occurs (a "passive" polluter). In any event, the determination of whether the pollution event qualifies as an "occurrence" basically will be an issue of fact rather than an issue that is resolved as a matter of law.

Trigger of Coverage. In construing CGL "occurrence" policies, courts uniformly hold that coverage is *triggered* under a specific policy when bodily injury or property damage takes place. Since an insurer will only be liable for indemnification or defense costs under its policy if the injury or damage takes place during the term of the policy, the determination of the date of the occurrence is important. Depending upon the "trigger of coverage" theory that a court adopts, a single policy or a number of policies may apply.

Determining which policy is "triggered" in a pollution case can be difficult. For example, a recently discovered pollution problem can be the result of a release of contaminants occurring many years earlier. Courts have adopted a number of theories in such cases to decide when an occurrence took place and, therefore, which policy applies. These include:

- The exposure theory (when the environment was first exposed to pollutants)
- The manifestation theory (when the pollution damage was, or should have been, first discovered)
- The injury-in-fact theory (when the release actually damaged the environment)
- The multiple-trigger or continuous trigger theory (all policies triggered from initial exposure to final manifestation)

All too often a policyholder encountering a pollution problem simply files a claim under its current CGL policy. But depending on the nature and timing of the release or releases that caused the problem and the trigger theory applied to the case, the current policy may not be applicable. The best strategy for a policyholder is to provide notice of the pollution problem to all insurers issuing policies covering the period from the earliest possible exposure to contaminants to the point of discovering the ultimate extent of the pollution.

Whether Cleanup Costs are Damages. Another issue involved in the question of what is covered by a CGL policy is whether cleanup costs are damages. The typical standard-form CGL policy provides:

> *The [insurer] will pay on behalf of the insured all sums which the insured shall become legally obligated to pay as damages because of (a) bodily injury or (b) property damage to which this insurance applies caused by an occurrence.*

This provision was drafted several years prior to enactment of CERCLA[4] and similar laws requiring payment of costs of response or remediation to clean up contaminated property. This non-traditional remedy does not fit neatly into the CGL policy's language and definitions that the policy insures only against damages. Insurers typically argue that environmental response and remediation costs are "equitable" in nature rather than being legal damages in the conventional sense, and, therefore, are not covered by CGL policies.

In early judicial decisions, it was held that CERCLA response costs are not covered damages. However, subsequent decisions overwhelmingly have been to the contrary. These later decisions generally reject any technical distinction between "equitable" and "legal" damages, and hold that money that the policyholder was legally obligated to pay as a result of property damage, including environmental harm, constitutes recoverable damages.

Exclusion of Property Owned or Leased by the Policyholder. CGL policies typically contain an express exclusion for damage to property: (1) owned or occupied by or rented by the policyholder, (2) used by the policyholder, or (3) in the care, custody, or control of the policyholder or to which the policyholder is for any purpose exercising physical control.

This exclusion has been held to preclude coverage for the costs of cleaning up the policyholder's own property where the contamination is situated solely on his property and does not pose a threat to property of a third party. However, the majority of courts do not interpret the exclusion to apply to coverage for the cost of pollution cleanups on a policyholder's own property if the cleanup is necessary to prevent the migration of pollutants into ground water or onto adjacent third-party property.

Cases involving groundwater contamination often revolve around who owns the ground water. If the state owns the ground water, the courts generally hold that contamination of it beneath a policyholder's property constitutes third-party damage. If the policyholder owns the ground water, some courts hold that contamination of it beneath a policyholder's property does not constitute third-party property damage unless and until it migrates beyond the policyholder's property line.

Another issue arising in the interpretation of the owned or leased-property exclusion is whether a "mere threat" of third-party property damage is enough to avoid the exclusion, or whether "actual" damage is required. The majority of courts examining the issue hold that costs incurred to eliminate or mitigate a threat to third-party property are not excluded. Other courts hold that actual injury to third-party property is required.

A more reasonable approach for resolving the "ownership" versus "non-ownership" and the "threat" versus "actual" damage issues was expressed in 1994 by the Seventh Circuit U.S. Court of Appeals in Patz v. St. Paul Fire & Marine Ins. Co.[5] The Patz court held that these issues were irrelevant because the policyholders were "not seeking to recover for damage to their own property," but rather, they were seeking "to recover the cost of the liability that the [state] imposed on them for maintaining a nuisance."

The Patz decision is consistent with the reasonable expectations of policyholders that they should not lose coverage because they acted prudently in mitigating on-site contamination in order to prevent off-site contamination. To hold otherwise effectively encourages policyholders to allow contaminants to migrate off-site in order to maximize insurance coverage at the expense of the environment.

"Sudden and Accidental" Pollution Exclusion. In about 1970, insurers, having become alarmed by the potential exposure from pollution coverage in CGL policies, developed a standardized pollution exclusion as follows:

> *It is agreed that the insurance does not apply to bodily injury or property damage arising out of the discharge, dispersal, release, or escape of smoke, vapors, soot, fumes, acids, alkalis, toxic chemicals, liquids or gases, waste materials or other irritants, contaminants or pollutants into or upon land, the atmosphere, or any water course or body of water; but this exclusion does not apply if such discharge, dispersal, release, or escape is sudden and accidental.*

Interpretation of the "sudden and accident" exception to the exclusion is probably the single most important legal issue confronting insurers and policyholders. Courts are deeply divided in their interpretations.

Regulatory History. In 1970, the Insurance Rating Board (IRB) and the Mutual Insurance Rating Board (MIRB), two of the national insurance industry rating boards, filed the standard-form pollution exclusion with various state insurance regulators for approval. The IRB and MIRB represented to state regulators that the exclusion was a mere clarification of existing coverage under standard occurrence-based liability policies. Policyholders contend that the IRB and MIRB made such representations so that the insurance companies could continue to charge the same level of premiums notwithstanding the exclusion.

In their filings, the IRB and the MIRB provided state regulators with the following official explanation:

> *Coverage for pollution or contamination is not provided in most cases under present policies because the damages can be said to be expected or intended and thus are excluded by the definition of occurrence. The [standard-form pollution] exclusion clarifies this situation so as to avoid any question of intent. Coverage is continued for pollution or contamination caused injuries when the pollution or contamination results from an accident . . .*

In examining the IRB's explanation of the 1970 pollution exclusion, the New Jersey Supreme Court in an important 1993 decision, Morton International, Inc. v. General Acc. Ins. Co., concluded that "[t]he first sentence of the IRB's explanation is simply not true."[6] As noted previously, the insurance industry drafted the 1966 occurrence-based general liability policy to cover property damage for gradual pollution, with no restriction on the suddenness of the pollution discharge. Indeed, as stated by one commentator, the occurrence-based policy "seemed tailor made to extend coverage to most pollution situations."[7]

The second sentence of the IRB's explanation, said the New Jersey Supreme Court, is even more misleading than the first one. Despite the IRB's assertion that the exclusion merely "clarifies this situation," the exclusion actually purports to preclude coverage for all coverage for unintentional pollution damage except for that caused by sudden and accidental discharges. In reviewing the second sentence of the explanation, the New Jersey Supreme Court observed that "[t]o describe a reduction of coverage of that magnitude as a 'clarification' not only is misleading, but comes perilously close to deception." Moreover, had the insurance industry acknowledged the true scope of this drastic reduction in coverage, state regulators would have been obligated to consider corresponding reductions in premiums.

The New Jersey Supreme Court further observed that final sentence of the IRB's explanation confuses matters even further. That "coverage is continued for pollution or contamination caused injuries when the pollution or contamination results from an accident" contradicts the very wording of the exclusion that coverage is not afforded unless the discharge of pollutants is "sudden and accidental." The explanation ignores this injuries/discharge distinction and "the conclusion is virtually inescapable that the memorandum's lack of clarity was deliberate."

Despite the insurance industry's 1970 representations to insurance regulators, insurers routinely deny policyholders' claims for gradual, accidental pollution. One thing is clear, the insurance industry either made false representations to insurance regulators in 1970 to induce regulators to approve the exclusion as a mere "clarification" of coverage without a reduction in premiums, or it made true representations to state regulators, and now insurers are denying coverage without a reasonable basis.

Policyholders' Interpretation. Policyholders contend that the term "sudden" is ambiguous, because it can mean either "abrupt" or "unintended." They argue that because of this ambiguity, "sudden and accidental" should be construed in their favor reasonably to mean "unexpected or unintended," i.e., that the exclusion precludes coverage only for pollution that was "unexpected and unintended" by the policyholder. A large number of courts have agreed with this interpretation.

Some courts ruling in favor of policyholders have held that the pollution exclusion is merely a restatement of the "occurrence" definition (i.e., that coverage is excluded only for pollution damage that was "unexpected and unintended" by the insured). Other courts, while agreeing that "sudden and accidental" means "unexpected and unintended," have concluded that the focus of the pollution exclusion is on the release of pollutants, and not on the resulting damage. This release/damage distinction is important in cases in which the policyholder intentionally discharged pollutants but did not intend or expect the resulting pollution damage.

Insurers' Interpretation. Insurers contend that the term "sudden and accidental" unambiguously precludes coverage for pollution caused by the gradual, expected, or intentional release of pollutants. A number of courts have agreed, holding that the term "sudden" has an ordinary temporal meaning (e.g., abrupt, over a short period of time) and does not include gradual or repeated releases of pollutants. Courts also have held that for coverage to exist, the release must be both "sudden" and "accidental," and that deliberate waste discharges are not "accidental" even if the insured did not intend any harm or the acts of a third party caused the discharge.

The "Relevant Discharge" Issue. As a result of the large number of courts that have determined that the term "sudden" does not preclude coverage for gradual pollution claims, the insurance industry has retreated to consideration of the "accidental" prong of the pollution exclusion's exception. The emerging issue is what is the "relevant" discharge, dispersal, release, or escape for applying the "sudden and accidental" exception?

Insurers argue that the initial discharge of waste into or upon "land" is the relevant discharge. Insurers contend that the deliberate placement of waste into land-based waste management units (e.g., landfills, pits, pond, lagoons), even if legally permitted and designed to contain the waste, cannot be characterized as "accidental." Therefore, insurers conclude that the resulting damages caused by "migration" of the wastes out of the unit are not covered under the "sudden and accidental" exception to the exclusion. Several courts agree.

Policyholders contend that the proper focus of the exclusion is not on the lawful disposal or placement of waste into a land-based waste management unit designed to contain the waste, but instead, is on the event or events causing third-party property damage and the policyholder's ultimate liability—the escape of waste from the unit into the surrounding environment. Furthermore, policyholders argue, until waste escapes from the containment unit causing third-party damages there can be no "occurrence," and, therefore, the pollution exclusion is not triggered. An emerging majority of courts agree.

The insurers' "relevant discharge" argument potentially results in an even larger reduction in coverage than their argument that "sudden" precludes coverage for all gradual pollution. When considered together, insurers effectively argue that the 1970 pollution exclusion eliminates coverage for most all lawful waste management practices. This argument is not only inconsistent with the insurance industry's representations to state regulators and policyholders' reasonable expectations of coverage, but it also approaches outright fraud.

The "Absolute" Pollution Exclusion. Responding to adverse decisions interpreting the 1970 pollution exclusion, insurance companies and their rating organizations began drafting a so-called "absolute" pollution exclusion purportedly to eliminate coverage for most pollution, including "sudden and accidental" pollution. The original 1985 standard-form version of the exclusion read that this insurance does not apply:

1. to "bodily injury" or "property damage" arising out of the actual, alleged, or threatened discharge, dispersal, release, or escape of pollutants:
   a. at or from premises owned, rented, or occupied by the named insured;
   b. at or from any site or location used by or for the named insured or others for the handling, storage, disposal, processing, or treatment of waste;
   c. which are at any time transported, handled, stored, treated, disposed of, or processed as waste by or for the named insured or any person or organization for whom the named insured may be legally responsible; or
   d. at or from any site or location on which the named insured or any contractors or subcontractors working directly or indirectly on behalf of the named insured are performing operations;
      i. if the pollutants are brought on or to the site or location in connection with such operations; or
      ii. if the operations are to test for, monitor, clean up, remove, contain, treat, detoxify, or neutralize the pollutants.
2. to any loss, cost, or expense arising out of any governmental direction or request that the named insured test for, monitor, clean up, remove, contain, treat, detoxify, or neutralize pollutants.

Pollutants means any solid, liquid, gaseous, or thermal irritant or contaminant, including smoke, vapor, soot, fumes, acids, alkalis, chemicals, and waste. Waste includes materials to be recycled, reconditioned, or reclaimed.

A number of states have approved the following endorsement subsequently modifying the "absolute" pollution exclusion:

> *Subparagraphs (a) and (d)(i) of paragraph (1) of this exclusion do not apply to "bodily injury" or "property damage" caused by heat, smoke, or fumes from a hostile*

fire. As used in this exclusion, a hostile fire means one that becomes uncontrollable or breaks out from where it was intended to be.

The terms "arising out of" and "pollutant" are so broad that their application is susceptible to overreaching by insurers (i.e., almost anything can be construed to "arise out of" the discharge, dispersal, release, or escape of a "pollutant").

In 1985, the Insurance Services Office, Inc. (ISO),[8] on behalf of its members and subscriber companies, filed the newly drafted exclusion for review and approval by state regulators in various states. In its filing, the ISO provided state regulators with the following official explanation:

This endorsement introduces a total pollution exclusion for bodily injury and property damage arising from the discharge of pollutants. The exclusion does not apply to damages arising out of products or completed operations nor to certain off-premises discharges of pollutants.

Importantly, the ISO's own explanation establishes that the exclusion was not as "absolute" or "total" as those words might suggest.

Subsequent to its approval of the 1985 pollution exclusion, the Texas State Board of Insurance held a hearing on the ISO's newly filed "claims-made" commercial general liability policy form. During the hearing, the insurance industry made additional representations in sworn testimony about the exclusion (insurers proposed that the 1985 pollution exclusion endorsement be incorporated into the proposed CGL form as a standard exclusion). Insurers admitted that the exclusion was overdrafted and ambiguous, but asserted that no one would apply it as written.

Despite the facts that representatives of the insurance industry testified under oath that the so-called "absolute" pollution exclusion was not really "absolute" and was overdrafted and ambiguous, that they testified that specific exclusions needed to be written for specific exposures, and that most policyholders had a different understanding of the meaning of "pollution" and would not understand that the new exclusion applied to them, the insurance industry failed to correct these problems.

As with its 1970 predecessor, the 1985 exclusion should not bar coverage in many situations involving environment impairment. The drafters of the policy language did not intend to "absolutely" exclude coverage for alleged "pollution," and courts should consider this important evidence in evaluating the 1985 pollution exclusion.

For the most part, courts, however, have found the 1985 exclusion and several similar versions unambiguously to preclude coverage for traditional industrial pollution liabilities, without regard to how or who caused the pollution.

Duty to Defend. With the rising costs of defending environmental claims, one of the most valuable portions of the CGL policy is the insurance company's duty to defend the policyholder in litigation. In most states, the duty to defend is determined solely from the face of the pleadings without reference to the truth or falsity of the allegations. The insurer is obligated to defend if the pleadings "potentially" state a case within the coverage of the policy.

An important issue that arises in the environmental context is whether an insurer's duty to defend is triggered when a policyholder receives a letter or other form of action (other than a traditional "suit") from an administrative agency, e.g., a PRP (Potentially Responsible Party) notice letter from the EPA (Environmental Protection Agency), or a notice of violation letter from a state agency. Although most courts will not require a traditional lawsuit to trigger the duty to defend, they are much more likely to find such a duty when the agency threatens formal action, particularly penalties.

Although the duty to defend is broader than the duty to indemnify, a finding by a court that the insurer does not have a duty to defend an administrative action does not relieve the insurer from its indemnity obligation. The duty to indemnify is a separate duty, without a "claim" or "suit" triggering requirement.

Policy Conditions. A liability insurer's duties to defend and indemnify its policyholder are conditioned upon the performance of certain duties that the insurance contract imposes on the policyholder. The policyholder's primary duties are to give notice of an accident or occurrence and to forward "suit papers" (e.g., claims, demands, or pleadings) to the insurer. The failure to satisfy either of these requirements can result in loss of coverage. The purpose of the notice requirements is to enable the insurer to investigate the circumstances of the occurrence while witnesses are still available and to gather all information necessary to respond to a claim or potential claim.

The standard-form CGL policy typically requires policyholders to notify the insurer of an occurrence "as soon as practicable" and suit papers must be forwarded to the insurer "immediately." Regardless of the precise form of the requirement, courts focus on the reasonableness of the notification to determine whether the policyholder has complied with the notice provision of the policy.

CGL policies also include requirements that the policyholder cooperate with the insurer and assist in making settlements, in the conduct of suits, and in enforcing any right of contribution or indemnity. In addition, CGL policies prohibit the policyholder from making voluntary payments or settling a claim without the insurer's knowledge and consent. A policyholder's failure to comply with these conditions can result in loss of coverage.

Personal Injury Coverage

Personal injury coverage can be provided to a policyholder for an additional premium as an endorsement to a CGL policy. This type of coverage is separate and apart from the coverage provided for "bodily injury" and "property damage."

The personal injury endorsement typically provides:

> *The Company will pay on behalf of the insured all sums that the insured shall become legally obligated to pay as damages because of injury (herein called "personal injury") sustained by a person or organization and arising out of one or more of the following offenses: . . . wrongful entry or eviction, or other invasion of the right of private occupancy.*

Policyholders contend that the term "personal injury" includes trespass, nuisance, and related property invasion claims. Under a trespass theory, policyholders characterize the release of contaminants onto or beneath a claimant's property as a "wrongful entry or eviction." They also argue that the release of contaminants interferes with the use and enjoyment of property, thus constituting a common law nuisance. This nuisance theory is aimed at establishing coverage under the "or other invasion of the right of private occupancy." These types of claims are embodied in many toxic tort suits as well as under CERCLA and similar state actions.

From the policyholder's perspective, the potential advantages to bringing environmental claims under the personal injury endorsement include: (1) avoidance of the "occurrence" trigger (instead coverage is provided for injury arising out of one of the enumerated "offenses"); (2) personal injury coverage typically contains a separate coverage grant and separate aggregate limits; (3) personal injury provisions cover intentional conduct; and (4) the pollution exclusion may not apply (i.e., (a) the personal injury endorsement has its own exclusions, which do not include a pollution exclusion, and (b) the pollution exclusion does not expressly preclude coverage for "personal injury"). Several courts have ruled in favor of policyholders on "personal injury" coverage for environmental claims.

Insurers counter that the personal injury endorsement was never intended to provide coverage for such environmental claims but rather was intended to cover torts that do not result in tangible injury, such as malicious prosecution, slander, or wrongful eviction, which were not already covered as "bodily injury" or "property damage." A slight majority of courts have denied "personal injury" coverage for environmental claims.

Although the case law is not well developed in this area, policyholders should not overlook the coverage that the personal injury endorsement potentially provides. Such coverage may be particularly valuable to policyholders who may have more difficulty obtaining coverage for "gradual" pollution claims under the traditional property damage/bodily injury portion of the policy because of the "occurrence" definition or the pollution exclusion.

Coverage Available in the Current Market

Obtaining pollution coverage in today's market is difficult, but not impossible. Few insurers are willing to write coverage on an occurrence basis. Instead, a number of companies sell separate liability policies specifically designed to cover pollution. These policies are commonly known as Environmental Impairment Liability (EIL) policies. Most EIL policies cover both "abrupt" and "gradual" pollution claims.

EIL policies generally are written on a "claims-made" basis, i.e., the pollution incident must occur, and a claim arising out of it must be made and reported to the insurer during the term of the policy. There is no coverage for a loss that occurs prior to the inception of the policy, even if a claim is made and reported during the policy term, or that occurs during the policy period but is not discovered and reported until after the policy expires. Most insurers, however, will sell the policyholder a "tail" that extends coverage for losses discovered and claims made and reported within a specified period of time after the policy is terminated.

Insurers often lead policyholders to believe that EIL coverage is much broader than is actually the case. For example, some policies appear to cover environmental cleanups of a policyholder's own property. In fact, coverage usually is limited to the cleanup of a third party's property. In addition, some policies require that pollution incidents be discovered and reported to the insurer within a brief time period, e.g., 72 hours. These limited reporting periods effectively eliminate coverage for difficult-to-discover pollution events, such as ground water contamination. Policyholders, therefore, should read the specimen policies carefully *before* purchasing EIL coverage to make sure that it meets their needs.

Insurers initially designed EIL coverage for businesses that treat, store, or dispose of hazardous waste. Today, however, EIL products are available for a number of other specific business operations. For examples, they are available for managing risks associated with petroleum storage tanks, asbestos and lead paint abatement, environmental remediation, real property transactions, and pesticide application.

Despite the increased availability of pollution liability coverage during the last few years, the market is still limited. Eventually the market may open up and provide broader pollution coverage, but this undoubtedly will come at a significant increase in premiums. The cost of minimizing the risks of pollution at the work site may be well worth the additional investment, given the uncertainty over the scope of pollution coverage under currently available insurance products.

Unless the United States Congress and the states drastically change the way they go about cleaning up our nation's hazardous waste sites, i.e., assessing liability without regard to fault, insured potentially responsible parties will continue to pursue insurance coverage in an attempt to shift the burden of paying the costs of environmental cleanups. Policyholders who neither expected nor intended the environmental damage at these sites should reasonably expect their insurance policies to provide coverage.

CHAPTER 24 ENDNOTES

1. The 1973 version of the standard-form CGL policy that the insurance industry promulgated further emphasized that an occurrence can take place over an extended period by redefining "occurrence" as follows:

> . . . an accident, including continuous or repeated exposure to conditions, which results in bodily injury or property damage neither expected nor intended by the insured.

2. Shell Oil Co. v. Winterthur Swiss Ins. Co., 15 Cal. Rptr. 2d 815 (Cal Dist. Ct. App.).
3. 27 F.3d 150, 152 (5th Cir.), reh'g denied en banc, 35 F.3d 564 (1994).
4. 42 USC §§ 9601–9662.
5. 15 F.3d 699 (7th Cir. 1994) (applying Wisconsin law).
6. 629 A.2d 831, 852 (1993).
7. E. Jahua Rosenkranz, Note, The Pollution Exclusion Through the Looking Glass, 74 Geo. L.J. 1237, 1251 (1986).
8. The Insurance Services Office, Inc. (ISO) is a trade organization that drafts standard policy language and files it for approval with state regulators. It is the successor to the Insurance Rating Board (IRB) and the Mutual Insurance Rating Board (MIRB).

APPENDIX 1
Environmental Acronyms and Other Terms

API	American Petroleum Institute
AQCR	Air Quality Control Region
ARB	California Air Resources Board
ASTM	American Society for Testing and Materials
BACT	Best Available Control Technology
BTEX	Benzene, Toluene, Ethylbenzene, and Xylene
CAA	Clean Air Act
CAAA 90	Clean Air Act Amendments of 1990
CCAA	California Clean Air Act of 1988
CERCLA	Comprehensive Environmental Response, Compensation, and Liability Act
CFR	Code of Federal Regulations
CGL	Comprehensive General Liability
CO	Carbon Monoxide
CWA	Clean Water Act
DEA	Diethanolamine
DNR	Louisiana Department of Natural Resources
DOG	Division of Oil and Gas and Geothermal Resources of the California Department of Conservation
DOT	Department of Transportation
DTSC	Department of Toxic Substances Control of the California Environmental Protection Agency
E&P	Exploration and Production
EPA	Environmental Protection Agency
FWS	Fish and Wildlife Service
GACT	Generally Available Control Technology
HAP	Hazardous Air Pollutant
HAZOP	Hazardous Operability Analysis
HWCA	California Hazardous Waste Control Act
H_2S	Hydrogen Sulfide
LAER	Lowest Achievable Emission Rate
MACT	Maximum Achievable Control Technology
MMS	Minerals Management Service

NAAQS	National Ambient Air Quality Standards
NGL	Natural Gas Liquids
NMFS	National Marine Fisheries Service
NO_2	Nitrogen Dioxide
NORM	Natural Occurring Radioactive Material
NOV	Notice of Violation
NO_x	Nitrogen Oxides
NPDES	National Pollution Discharge Elimination System
NPL	National Priorities List
NSPS	New Source Performance Standard
NSR	New Source Review
O_3	Ozone
OCS	Outer Continental Shelf
O&G	Oil and Grease
OPA or OPA 90	Oil Pollution Act of 1990
OSHA	Occupational Safety and Health Act (or Administration)
PCB	Polychlorinated Biphenyl
POTW	California Publicly-Owned Treatment Works
PM-10	Particulate Matter of 10 Microns in Diameter or Less
PRP	Potentially Responsible Party
ppm	Parts Per Million
PSD	Prevention of Significant Deterioration
RACT	Reasonably Available Control Technology
RCRA	Resource Conservation and Recovery Act
RMP	Risk Management Program (or Plan)
RRC	Railroad Commission of Texas
SARA	Superfund Amendments and Reauthorization Act of 1986
SIP	State Implementation Plan
SO_2	Sulfur Dioxide
SPCC	Spill Prevention, Control, and Countermeasure
SWDA	Solid Waste Disposal Act
TDH	Texas Department of Health
THC	Total Hydrocarbons
TPCA	California Toxic Pits Cleanup Act
TPH	Total Petroleum Hydrocarbons
tpy	Tons Per Year

TSCA	Toxic Substance Control Act
TSD	Technical Support Document
TSDF	Treatment, Storage and Disposal Facility
USC	United States Code
VOC	Volatile Organic Compound

APPENDIX 2
Statutory Extracts from CERCLA

APPENDIX 2–1
PRPs

CERCLA S*ec.* 107(*a*) **(42 USC 9607)**

Notwithstanding any other provision or rule of law, and subject only to the defenses set forth in subsection (b) of this section—

1. the [present] *owner and [or] operator of a vessel or a facility,*
2. any person who *at the time of disposal* of any hazardous substance *owned* or operated any facility at which such hazardous substances were disposed of,
3. any person who by contract, agreement, or otherwise *arranged for disposal* or treatment, or arranged with a transporter for transport for disposal or treatment, of hazardous substance owned or possessed by such persons, by any other party or entity, at any facility or incineration vessel owned or operated by another party or entity and containing such hazardous substance, and
4. any person who *accepts* or accepted any hazardous substances *for transport* to disposal or treatment facilities, incineration vessels or sites selected by such person, *from which there is a release* (or a threatened release which causes the incurrence of response costs), *of a hazardous substance, shall be liable* for—

A. all costs of removal or remedial action incurred by the United States Government or a State or an Indian tribe *not inconsistent* with the national contingency plan;

B. any other necessary costs of response incurred by any other person *consistent* with the national contingency plan;
C. damages for injury to, destruction of, or loss of *natural resources,* including the reasonable costs of assessing such injury, destruction, or loss resulting from such a release; and
D. the costs of any *health assessment* or health effects study carried out under section 104(i).

APPENDIX 2–2
Facility

CERCLA *Sec.* 101(9) **(42 USC 9601)**

The term "facility" means (A) any building, structure, installation, equipment, pipe or pipeline (including any pipe into a sewer or publicly owned treatment works), well, pit, pond, lagoon, impoundment, ditch, landfill, storage container, motor vehicle, rolling stock, or aircraft, or (B) any *site or area* where a hazardous substance has been deposited, stored, disposed of, or placed, or *otherwise come to be located;* but does not include any consumer product in consumer use or any vessel.

APPENDIX 2–3
Disposal

SWDA *Sec.* 1004(3) **(42 USC 6903)**

The term “disposal” means the discharge, deposit, injection, dumping, spilling, leaking, or placing of any solid waste or hazardous waste *into or on any land or water* so that such solid waste or hazardous waste or any constituent thereof *may* enter the environment or be emitted into the air or discharged into any waters, including ground waters.

APPENDIX 2–4
Release

CERCLA *Sec.* 101(22) **(42 USC 9601)**

The term "release" means any spilling, leaking, pumping, pouring, emitting, emptying, discharging, injecting, escaping, leaching, dumping, or disposing into the environment (including the abandonment or discharge of barrels, containers and other closed receptacles containing any hazardous substance or pollutant or contaminant), . . .

APPENDIX 2–5
Third Party Defense

CERCLA *Sec.* 107(*b*) (42 USC 9607)

There shall be no liability under subsection (a) of this section for a person otherwise liable who can establish by a preponderance of the evidence that the release or threat of release of a hazardous substance and the damages resulting therefrom were caused *solely* by-—

1. an act of *God;*
2. an act of *war;*
3. an act or omission of a *third party other than* an *employee* or *agent* of the defendant, or than one whose act or omission occurs in connection with a *contractual relationship,* existing directly or indirectly, with the defendant (except where the sole contractual arrangement arises from a published tariff and acceptance for carriage by a common carrier by rail), *if* the defendant establishes by a preponderance of the evidence that (a) he exercised *due care* with respect to the hazardous substance concerned, taking into consideration the characteristics of such hazardous substance, in light of all relevant facts and circumstances, *and* (b) he took *precautions* against foreseeable acts or omissions of any such third party and the consequences that could foreseeably result from such acts or omissions; or
4. any combination of the foregoing paragraphs.

APPENDIX 2–6
Innocent Landowner

CERCLA *Sec.* 101(35) (42 USC 9601)

(35)(A) The term "contractual relationship", for the purpose of section 107(b)(3), includes, but is not limited to, land contracts, deeds or other *instruments transferring title or possession,* unless the real property on which the facility concerned is located was *acquired* by the defendant *after the disposal* or placement of the hazardous substance on, in, or at the facility, *and* one or more of the circumstances described in clause (i), (ii), or (iii) is also *established by the defendant by a preponderance of the evidence:*

i. At the time the defendant acquired the facility the defendant *did not know* and *had no reason to know* that any hazardous substance which is the subject of the release or threatened release was disposed of on, in, or at the facility.

ii. The defendant is a government entity which acquired the facility by escheat, or through any other involuntary transfer or acquisition, or through the exercise of eminent domain authority by purchase or condemnation.

iii. The defendant acquired the facility by inheritance or bequest.

In addition to establishing the foregoing, the defendant must establish that he has satisfied the requirements of section 107(b)(3) (a) and (b).[1]

(B) To establish that the defendant had no reason to know, as provided in clause (i) of subparagraph (A) of this paragraph, the defendant must have undertaken, at the time of acquisition, *all appropriate inquiry* into the previous *ownership* and *uses* of the property consistent with good commercial or customary practice in an effort to minimize liability. For purposes of the preceding

[1]So in law. Should be 107(b)(3) (A) and (B).

sentence the court shall take into account any specialized knowledge or experience on the part of the defendant, the relationship of the purchase price to the value of the property if uncontaminated, commonly known or reasonably ascertainable information about the property, the obviousness of the presence or likely presence of contamination at the property, and the ability to detect such contamination by appropriate inspection.

(C) Nothing in this paragraph or in section 107(b)(3) shall diminish the liability of any previous owner or operator of such facility who would otherwise be liable under this Act. Notwithstanding this paragraph, if the defendant obtained actual knowledge of the release or threatened release of a hazardous substance at such facility when the defendant owned the real property and then subsequently *transferred ownership* of the property to another person *without disclosing* such knowledge, such defendant shall be treated as liable under section 107(a)(1) and *no defense* under section 107(b)(3) shall be available to such defendant.

(D) Nothing in this paragraph shall affect the liability under this act of a defendant who, by any act or omission, caused or *contributed* to the release or threatened release of a hazardous substance, which is the subject of the action relating to the facility.

APPENDIX 2–7
Secured Creditor Exemption

CERCLA Sec. 101(20)A **(42 USC 9601)**

The term “owner or operator” . . . does not include a person, who, without participating in the management of a . . . facility, holds indicia of ownership primarily to protect his security interest in the . . . facility.

APPENDIX 3

Final General NPDES *Permits* Oil *and* Gas *Extraction Point Source Category* (*Exclusive of* Offshore *Alaska*)

- Offshore Southern California. 48 Fed. Reg. 55031 (Dec. 8, 1983).
- Outer Continental Shelf Gulf of Mexico. 51 Fed. Reg. 21897 (July 9, 1986).
- Onshore Louisiana, New Mexico, Oklahoma, and Texas. 56 Fed. Reg. 7698 (Feb. 25, 1991).
- Coastal Waters Louisiana and Texas. 60 Fed. Reg. 2387 (Jan. 9, 1995) and 60 Fed. Reg. 12555 (March 7, 1995 Modification).

APPENDIX 4

Guidance on Clean Water Act Definition of Oil (Oil of Any Kind, In Any Form)

OIL

Cutting oils
Transformer oils
#2–#6 Fuel oils, bunker fuel
Mix of benzene, toluene, and xylene
Jet fuels (JP–4 etc.)
Kerosene
Diesel fuel
Gasoline
Motor oils and lubricating oils
Animal and vegetable fats/oils
  Whale oil
  Coconut oil
  Corn oil
  Tallow
  Meat rendering
  Fish oil
  Tung oil
  Linseed oil
  Cottonseed oil
Fuel oil sludge
Waste oil
Oil mixtures
Paraffin wax
Asphalts
Tars (petroleum)
Crude oils
Naphtha
Mineral spirits
Refinery petroleum waste
Lacquer base paints/varnishes
High-temp hearing oils (Dowthern, Therminol, etc.)

NOT AN OIL*

Toluene
Molasses
Plating waste
Stearic acid
Cresol
Hexanol
Methyl mercury
DDT
Natural gas condensate
Naphthene
Phenols
Toxaphene
Butanol
Ethanol
Methanol
TBA, MTBE (gasoline additives)
Benzene
Ketones
Acetone
Propanol
Glycol
Xylene
Ethylene/propylene Glycol (antifreeze)
Coal tar

**Note 1:* To determine if it is an oil, the following three conditions must be met:

1. It is an isomer. An isomer is a compound that is identical to another compound in its formula but different in its molecular arrangement. An example is xylene, which has 3 isomers: ortho-xylene, meta-xylene, and para-xylene. The formula $C_6H_4(CH_3)_2$ is the same for all of these isomers. The difference in appearance is due to the different placement of the methyl groups on the benzene ring.
2. By itself, the isomer is a non-oil. However, when combined with 3 or more in a homologous series it becomes an oil. A homologous series is a series of compounds in which each successive member has one more CH_2 group in its molecule than the preceding member. In the example of xylene, the homologous series is: benzene C_6H_8, toluene C_7H_8, and xylene C_9H_{10}.
3. Any non-petroleum derived material that is extractable by an organic solvent qualifies as an oil providing that its chemical structure is not defined or the substance is a complex mixture of chemical compounds.

Note 2: For additional information on differentiating oil and hazardous substances, refer to:

Oil versus Other Hazardous Substances. C. Hugh Thompson, Water Quality Office, Environmental Protection Agency.

Properties and Effects of Non-Petroleum Oils. Hans J. Crump & Allen L. Jennings, Environmental Protection Agency, Washington D.C.

**Both of these papers were from the proceedings of the 1971 Conference on Prevention and Control of Oil Spills.

1975 Letter to National Renderers Association. Kenneth E. Biglane, HQ. US EPA, Director for Oil and Special Materials Control Division.

Note 3: This list is *not* all inclusive, but rather identifies some common substances encountered at different facilities.

APPENDIX 5

Contents of Spill Reports Required for Facilities Subject to SPCC Plan Requirements

Source: 40 C.F.R. *ff* 112.4(*a*).

If a facility subject to SPCC plan requirements discharges more than 1,000 gallons of oil into "navigable waters" in any one spill event, or has two or more discharges of "harmful quantities," i.e., quantities that produce a sheen, within any 12-month period, then the owner or operator must prepare and submit to the EPA regional administrator for the region in which the facility is located a detailed report within 60 days, with a copy to the state agency with jurisdiction over oil and gas-related water pollution control activities. The report must include:

- Name of the facility
- Name(s) of the owner or operator of the facility
- Location of the facility
- Date and year of initial facility operation
- Maximum storage or handling capacity of the facility and normal daily throughput
- Description of the facility including maps, flow diagrams, and topographical maps
- A complete copy of the SPCC Plan with any amendments
- The causes(s) of such spill, including a failure analysis of the system or subsystem in which the failure occurred
- The corrective actions and/or countermeasures taken, including an adequate description of equipment repairs and/or replacements
- Additional preventive measures taken or contemplated to minimize the possibility of recurrence
- Such other information as the Regional Administrator may reasonably require pertinent to the Plan or spill event

APPENDIX 6

SPCC *Compliance Inspection Report* (EPA *Region* 6)

UNITED STATES ENVIRONMENTAL PROTECTION AGENCY

REGION 6
1445 ROSS AVENUE, SUITE 1200
DALLAS TX, 75202-2733

SPCC COMPLIANCE INSPECTION REPORT

DOCKET: ____________________ INSPECTION DATE: ________________

USEPA INSPECTORS: ______________________ and ____________________________

Inspection Type: - Announced - Unannounced - Amendment - Follow up

Facility in operation? Y N Applicable to 40 CFR; part 112? Y N

OPERATOR: _____________________________ TELEPHONE: _____________________
==========
ON SCENE CONTACT: _______________________ TITLE: _________________________

NAME OF BUSINESS: ____

HOME BUSINESS ADDRESS: ____

CITY: __________________________ STATE: _________________ ZIP: ___________

FACILITY/LEASE: ____
===============
SECTION: _______ TOWNSHIP: ______________ RANGE: _________ OTHER: _________

CITY: __________________________ COUNTY: ________________ STATE: _________

Total storage capacity: ________________ bbls Oil type: ______________________

Number of tanks: _________________________ Size of tanks: ____________________

Nearest body of water (distance): ____

DEFICIENCIES:

( ) No secondary containment
( ) Inadequate secondary containment
( ) Containment not sufficiently impervious
( ) Pooling liquid within containment
( ) Inadequate integrity of storage tanks
( ) Inadequate integrity of flowlines
( ) Inadequate freeboard of pits
( ) No drip pans on off-load lines
( ) No SPCC Plan at facility
( ) SPCC Plan not implemented as per 112.7
( ) SPCC Plan inadequate as per 112.7
( ) Other _________________________

( ) Please submit a copy of the SPCC Plan to U.S. EPA

CERTIFIED ENGINEER: ______________________________ NUMBER: _______________

DATE SPCC PLAN CERTIFIED: ____

APPENDIX 7
Exploration, Development, or Production Wastes Exempt from RCRA Subtitle C

Source: 53 Fed. Reg. 25446, 25453–54 (July 6, 1988)

- Produced water
- Drilling fluids
- Drill cuttings
- Rigwash
- Drilling fluids and cuttings from offshore operations disposed of onshore
- Well completion, treatment, and stimulation fluids
- Basic sediment and water, and other tank bottoms from storage facilities that hold products and exempt waste
- Accumulated materials such as hydrocarbons, solids, sand, and emulsions from production separators, fluid treating vessels, and production impoundments
- Pit sludges and contaminated bottoms from storage or disposal of exempt wastes
- Workover wastes
- Gas plant dehydration wastes, including glycol-based compounds, glycol, filter media, backwash, and molecular sieves
- Gas plant sweetening wastes from sulfur removal, including amines, amine filters, amine filter media, backwash, precipitated amine sludge, iron sponge, and hydrogen sulfide scrubber liquid and sludge
- Cooling tower blowdowns
- Spent filters, filter media, and backwash (assuming the filter itself is not hazardous and the residue in it is from an exempt waste stream)

- Packing fluids
- Produced sand
- Pipe scale, hydrocarbon solids, hydrates, and other deposits removed from piping and equipment prior to transportation
- Hydrocarbon-bearing soil
- Pigging wastes from gathering lines
- Wastes from subsurface gas storage and retrieval, except for the nonexempt wastes listed in Appendix 8
- Constituents removed from produced water before it is injected or otherwise disposed of
- Liquid hydrocarbons removed from the production stream but not from oil refining
- Gases removed from the production stream, such as hydrogen sulfide, carbon dioxide, and volatilized hydrocarbons
- Materials ejected from a producing well during the process known as blowdown
- Waste crude oil from primary field operations and production
- Light organics volatilized from exempt wastes in reserve pits, impoundments, or production equipment

APPENDIX 8

Exploration, Development, or Production Wastes Nonexempt from RCRA Subtitle C

Source: 53 Fed. Reg. 25446, 25454 (July 6, 1988)

- Unused fracturing fluids or acids
- Gas plant cooling tower cleaning wastes
- Painting wastes
- Oil and gas service company wastes, such as empty drums, drum rinsate, vacuum truck rinsate, sandblast media, painting wastes, spent solvents, spilled chemicals, and waste acids
- Vacuum truck and drum rinsate from trucks and drums transporting or containing non-exempt waste
- Refining wastes
- Liquid and solid wastes generated by crude oil and tank bottom reclaimers
- Used equipment lubrication oils
- Waste compressor oil, filters, and blowdown
- Used hydraulic fluids
- Waste solvents
- Waste in transportation pipeline-related pits
- Caustic or acid cleaners
- Boiler cleaning wastes
- Boiler refractory bricks
- Boiler scrubber, fluids, sludges, and ash
- Incinerator ash

- Laboratory wastes
- Sanitary wastes
- Pesticide wastes
- Radioactive tracer wastes
- Drums, insulation, and miscellaneous solids

APPENDIX 9
Clean Air Act Detailed Summary

Introduction

The Clean Air Act of 1990 (CAA) has evolved through several statutes.[1] Its stated purposes are "to protect and enhance the quality of the Nation's air resources so as to promote the public health and welfare and the productive capacity of its population;"[2] "to initiate and accelerate a national research development program to achieve the prevention and control of air pollution;"[3] "to provide technical and financial assistance to State and local governments; . . ."[4] and "to encourage and assist the development and operation of regional air pollution prevention and control programs."[5] The primary goal is "to encourage or otherwise promote reasonable Federal, State, and local governmental actions, consistent with" the CAA.[6]

Oil and gas exploration and production (E&P) is affected by the CAA as it is implemented by the EPA and the state regulatory bodies. The CAA is a complex piece of legislation embodying many programs, structures, and mechanisms that impose requirements upon the industry and how it operates. The basic structure of the CAA as it affects E&P is described in this appendix.

Structure and Mechanisms of the Clean Air Act

To comply with the CAA, an understanding of its structure and the mechanisms it employs to achieve its purpose is needed. The four statutory techniques used by the CAA are:

- A harm-based ambient quality regulatory approach controlling the most commonly produced and significant air pollutants. This technique is created by CAA Sections 107, 108, 109, and 110.
- Automobile and truck tailpipe emission standards given in Title II of the CAA.
- A technology-based standard approach established in Section 111 of the CAA.
- Regulation of hazardous air pollutants (HAPs) by a technology-based regulation strategy.[7,8]

All of these, except automobile tailpipe emissions standards, affect E&P.

The CAA relies on several programs and mechanisms to achieve and maintain air quality: a) national ambient air quality standard (NAAQS), b) air quality control regions (AQCRs), c) state implementation plans (SIPs), d) new source performance standards (NSPS), e) hazardous air pollutants (HAPs), and f) technology performance standards (BACT, RACT, MACT, LAER). Finally, permit requirements are established to govern the release of regulated substances into the atmosphere.

NAAQS

CAA Section 108 requires the EPA administrator to list criteria air pollutants.[9,10] Statutory standards for the list are: (1) it "cause[s] or contribute[s] to air pollution which may reasonably be anticipated to endanger public health or welfare;" and (2) "the presence of which in the ambient air results from numerous or diverse mobile or stationary sources".[11] Once listed, the administrator must publish air quality criteria for that pollutant; the document will cite the scientific knowledge on the identifiable effects on public health and welfare.[12] He must also publish information on emission control techniques.[13] The issuance of air quality criteria and air pollution control techniques must be announced in the Federal Register.[14]

Simultaneously, the administrator must also publish regulations proposing both a national primary and secondary ambient air quality standard.[15] These are referred to as "NAAQS".[16] The primary NAAQS are health based and must provide an adequate margin of safety to protect the public health,[17] while the secondary NAAQS are to protect the public welfare.[18,19]

Primary and secondary NAAQS are found at 40 CFR, Part 50. An air pollutant includes any precursors to the formation of any air pollutant to the extent identified by the EPA Administrator.[20]

State Implementation Plan (SIP)

The SIP is required by CAA Sections 108 and 110. CAA Section 110 requires each state to submit for primary and secondary NAAQS a plan for implementation, maintenance, and enforcement of the primary standard for each air quality control region (AQCR), which is discussed below.[21] The SIP must be adopted after public hearings,[22] and must meet the following requirements:[23]

- Include enforceable emission limitations and other control measures;
- Acquire air quality data;
- Provide an enforcement program for emission limitations, and the regulation of the modification and construction of any stationary source including a permit program;
- Contain provisions to control interstate impacts;
- Contain adequate assurances of personnel, funding, and authority to carry out the SIP;
- Require monitoring and emission data for stationary sources;
- Provide for emergency plans in the event of an imminent and substantial danger;
- Provide for revision of the SIP;

- Meet requirements of CAA Section 172(c) for nonattainment areas;
- Meet requirements for prevention of significant deterioration and visibility;[24]
- Provide for air quality modeling;
- Require fees for permits; and
- Provide for consultation and participation by local government.

Additional requirements for the SIP are in the PSD,[25] Nonattainment,[26] and HAPs[27] provisions of the CAA.

Air Quality Control Regions

CAA Section 107(b) requires the establishment of air quality control regions (AQCRs) and either reaffirms existing regions or calls for designation of regions after passage of the CAA.[28] The AQCR is a region where the meteorological, social, economic, and political considerations make it a "logical" geographical area to be regulated.[29] The EPA Administrator has the authority to designate as an AQCR any interstate area or major intrastate area that he deems necessary and appropriate for attaining and maintaining the NAAQS.[30]

CAA Section 107 gives each state the primary responsibility for assuring its air quality and requires it to submit a SIP that specifies the manner in which the NAAQS will be achieved within each region of the state.[31] Each state must then submit for each criteria pollutant within each AQCR a designation of whether that AQCR is: (1) nonattainment, (2) attainment, or (3) unclassifiable.[32] Unless otherwise provided, the SIP must provide for achieving the NAAQS for the AQCR within 5 years after being designated as nonattainment.[33]

Attainment Designation. If an AQCR is designated as attainment, it nevertheless must meet a requirement to maintain the air quality.[34] A program was established and then codified in the CAA to "prevent significant deterioration" (PSD) of the air quality.[35] A source is a "major stationary source" or a "major emitting facility" under the CAA if it is a stationary source of air pollutants or a facility that directly emits, or has the potential to emit, 100 tons per year of any air pollutant.[36] This general definition is supplemented by the definition in the PSD part of the CAA, which provides that a "major emitting facility" is one of 28 source categories that emit or have the potential to emit 100 tons per year and one of the categories is sulfur recovery plants.[37]

A PSD program contains several requirements:

(1) Areas are classified as Class I (national parks), Class II (all other areas previously designated as attainment or as unclassifiable), and Class III.[38] Based upon the class, an AQCR may be degraded by increments from its baseline air quality, without losing its designation. For example, on an annual basis: a Class I area can have a sulfur dioxide increase of 2 micrograms per cubic meter,[39] a Class II area can have a sulfur dioxide increase of 20 micrograms per cubic meter,[40] and a Class III area can have a sulfur dioxide increase of 40 micrograms per cubic meter.[41] The CAA also specifies increments for shorter times for sulfur dioxide. The act provides for the EPA to develop a PSD program to deal with other NAAQS pollutants.[42] The increments allowed under PSD, however, cannot cause the air quality in the AQCR to exceed either the primary or secondary NAAQS.[43]

(2) Preconstruction Review Program is established for all "major emitting facilities"[44] and requires that all such facilities use a stringent level of control (BACT).[45]

(3) Visibility protection for Class I areas[46] (applicable if E&P operations will affect national park or wilderness area).

Nonattainment Designation. If an area does not meet the NAAQS, it is classified as nonattainment. The SIP must provide for attaining the ambient air quality standards within 5 years,[47] unless otherwise provided by the CAA.[48] The CAA provides for a permit[49] and has the following extensive requirements for a SIP nonattainment area:[50] (1) implementation of all reasonably available control measures (including existing sources),(2) reasonable further progress towards attainment, (3) an inventory of actual emissions from all sources, (4) identification and qualification of the new emissions to be allowed from construction and operation of major new or modified sources, (5) permit program for new or modified major stationary sources, and (6) other measures such as enforceable emission limitations and timetables for compliance.

To obtain a permit, the applicant must obtain sufficient offsetting emissions. The total allowable emissions from existing sources in the region, from new or modified sources that are not major emitters, and from the proposed source must be sufficiently less than total emissions from existing sources prior to the permit application. This must represent reasonable further progress, that is, there is a net decrease in emissions[51] or the new or modified major stationary source is located in an economic development zone.[52] The proposed source must use LAER controls.[53] The owner must demonstrate that all major stationary sources owned or operated by such a person in such a state are subject to emission limitation and are in compliance.[54] Further, it must be shown that the benefits of the proposed source significantly outweigh the environmental and social costs imposed as a result of its location, construction, or modification after an analysis of alternative sites.[55] Offsets are required by obtaining emissions reduction of such air pollutant from the same source or other sources in the same nonattainment area.[56]

If a SIP is inadequate or not being implemented, if a required submission from the state to the EPA is not made, or if any required submission of an SIP is disapproved by the EPA,[57] then sanctions are available to the EPA Administrator. These include withholding of highway funds and ordering the emission offsets required for a new or modified source to be twice the increase in emissions from the source.[58]

Ozone nonattainment requirements. Areas are classified as Marginal, Moderate, Serious, Severe, and Extreme in their nonattainment of the ozone NAAQS. Requirements are cumulative from Marginal to Extreme, i.e., each area in a given classification has special requirements, which it must meet for all lesser classifications. Perhaps the major impact is the changing definition of major source. The general definition of a major source is one that emits 100 tons per year of a regulated pollutant;[59] however, this definition is modified for various parts of the CAA. The emission level for a major source of NO_x and VOCs is set generally as follows for nonattainment areas:

Area Classification	Major Source tons/yr.	Major Modified Source tons/yr.	Minimum Offset Ratio
Marginal[60]	100	40	1.1:1[61]
Moderate[62]	100	40	1.15:1[63]
Serious[64]	50	25	1.2:1[65]
Severe[66]	25	25	1.3:1[67]
Extreme[68]	10	0	1.5:1[69]

Unclassified. An AQCR is considered unclassifiable only if there is no data that can be used. However, since the requirements for a

SIP include continuous monitoring of air quality, data will ultimately become available. If an area is unclassifiable, then it is regulated as if it were an attainment area.[70]

NSPS

The New Source Performance Standard (NSPS) control technique is found in CAA Section 111 "Standards of performance for new stationary sources."[71] A "standard of performance" is a "standard for emissions of air pollutants which reflects the degree of emission limitation achievable through the application of the best system of emission reduction which (taking into account the cost of achieving such reduction and any nonair quality health and environmental impact and energy requirements) the Administrator determines has been adequately demonstrated."[72] It is a federal standard for new sources,[73] which include any stationary sources whose construction or modification begins after the publication of applicable regulations.[74] A stationary source is any building, structure, facility, or installation that emits or may emit any air pollutant.[75] The Administrator publishes a list of categories of stationary sources that contribute significantly to air pollution and may reasonably be anticipated to endanger the public health and welfare.[76] It then publishes standards of performance within one year of listing a source and reviews such standards at least once every 8 years.[77] The Administrator makes regulations under which states must submit a plan to establish standards of performance for existing sources for any air pollutant.[78]

It is unlawful for the owner or operator of any new source to operate it in violation of any applicable NSPS.[79] If a NSPS is not feasible, then the Administrator may prescribe a design, equipment, work practice, or operational standard, or combination of these that reflects the best technological system of continuous emission reduction.[80]

NSPSs are published by the EPA in the CFR[81] and are updated in the Federal Register. The EPA also publishes Control Technique Guidelines, which can be located by searching the National Technical Information Service literature database or DIALOG.

Hazardous Air Pollutants (HAPs)

The CAA continues and expands the regulation of pollutants that are considered HAPs.[82] A HAP[83] is any one of the 189 substances listed in CAA Section 112(b),[84] and any substance that the Administrator determines "present, or which may be present, through inhalation or other routes of exposure, a threat of adverse human health effects."[85] A major source is any stationary source or group of sources (located within a contiguous area and under common control) that emits or has the potential to emit 10 tons per year of any single HAP or 25 tons per year of HAPs in the aggregate.[86] An area source is any stationary source of HAPs that is not a major source.[87] The Administrator publishes a list of all categories and subcategories of sources[88] and establishes emission standards.[89]

Emission standards are promulgated by the EPA Administrator for each listed category and subcategory of sources of HAPs.[90] Such standards are applicable to new and existing sources and require the maximum degree of reduction (MACT) in emissions of HAPs, taking into account the cost and any non-air quality health and environmental impacts and energy requirements.[91] The reduction is to be achieved through measures including, but not limited to, those that: (1) reduce the volume or eliminate the emission; (2) enclose the systems or processes; (3) collect, capture, or treat the emissions; (4) are design, equipment, work practice, or operational standards; or (5) are a combination of the preceding.[92] The standards must have an ample margin of safety[93] and no MACT standard shall be interpreted so as to diminish or replace the requirements of an otherwise applicable more stringent emission limitation (NSPS, PSD, nonattainment, or other CAA requirement, or a standard issued under state authority)[94] For new sources, MACT must be at least as stringent as the emission control achieved in practice by the best controlled similar source (as determined by the Administrator).[95] For an existing source, MACT is not required to be as stringent as for a new source but it must not be less stringent than: (1) the average emission limitation achieved by the best 12% of existing sources (if there are 30 or more such sources) or (2) the average emission limitation achieved by the best performing five (5) sources (if there are less than 30 such sources).[96] MACT is required for a new source or any modification of a major source of HAPs.[97] The provisions of the PSD program do not apply to pollutants listed as HAPs;[98] but many HAPs are also volatile organic compounds and are regulated as precursors to ozone.

Risk Management Program. A general duty to identify hazards, to design and maintain a safe facility, and to minimize the consequences of accidental releases is imposed on owners and operators of facilities handling hazardous substances listed in 112(r)(3) or any other extremely hazardous substance.[99] This section, however, is not to be used to create any civil liability for personal injury or damage to property of others.[100] The listing of a substance is made considering the: (1) severity of any acute health effects associated with a release, (2) likelihood of an accidental release, and (3) potential magnitude of human exposure to the accidental release.[101] A threshold quantity must be established by the EPA Administrator that accounts for the toxicity, reactivity, volatility, dispersability, combustibility, or flammability of the substance and for the amount of the substance that as a result of an accidental release is known (or may reasonably be anticipated) to cause death, serious injury, or serious adverse health effects.[102] The proposed rule was published[103] January 19, 1993, and it listed 100 acutely toxic substances, 62 flammable substances, and high explosives as a class.[104] A final rule was published January 31,1994[105] stating that if a facility has more than the threshold quantity, then it must have a risk management plan.[106] The plan must include: (1) a hazard assessment to assess the potential effects of an accidental release (including downwind effects), (2) a program to prevent accidental release (including safety precautions, maintenance, monitoring, and training) and (3) a response program that identifies specific actions to be taken in the event of an accidental release to protect human health and the environment, including notification of local authorities.[107] The owner or operator must also register the risk management plan with the EPA Administrator, the Chemical Safety and Hazard Investigation Board, the state, and any local agency/entity having emergency response responsibility for an accidental release.[108] The proposed rule on the risk management program was published October 20, 1993 at 58 FR 54190 and it contains extensive discussion and detailed requirements to meet the mandates for the risk management program of CAA Section 112(r).

Technologies—BACT, RACT, MACT, LAER

The Administrator must establish a central database and include all control technology information received pursuant to state actions requiring permits for sources.[109] The control technologies have been introduced above in the various programs of the CAA. They are defined as follows:

- BACT, or Best Available Control Technology, is an emission limitation based on the maximum degree of reduction of each regulated pollutant from any major emitting facility. The permitting authority, on a case-by-case basis, determines what is achievable for the facility through application of production processes, and available methods, systems, and techniques. The permitting authority must consider energy, environmental, and economic impacts and other costs. In no event can BACT in an attainment area allow emissions that will exceed the standards set in Section 7411 (NSPS) or Section 7412 (HAPS).[110]

- RACT, or Reasonably Available Control Technology, is referred to in Section 7502(c)(1) (nonattainment) but there is no definition.
- LAER, or Lowest Achievable Emission Rate, means that rate of emissions that is the most stringent emission limitation contained in the SIP of any state for such class or category of source unless it can be demonstrated that it is not achievable, or the most stringent emission limitation that is achieved in practice by that class or category of source.[111] It cannot be less stringent than the applicable NSPS.
- MACT, or Maximum Achievable Control Technology, is required of new or existing sources. It is the maximum degree of reduction in emissions that the EPA Administrator determines is achievable for new or existing sources in the same category. Such determination will consider cost of achieving such a reduction, any non-air health and environmental impacts, and the energy required.[112] It cannot be less stringent than the emission controlled that is achieved in practice by the best controlled similar source.[113]

The NSPSs are a technology floor for all new sources, that is, the control technology used cannot be less stringent than the corresponding NSPS.

Technology Standard	Program
RACT	Nonattainment—existing sources
BACT	PSD—new/modified sources
LAER	Nonattainment—new/modified sources
MACT	Hazardous Air Pollutants—new or existing sources
NSPS	New/modified sources[114]

Permits

Preconstruction permits under new source reviews are required for attainment (PSD) and nonattainment areas. Also, now required are operating permits covered in Title 5 of the CAA Amendments of 1990.[115] The operating permit is required for all sources[116] and it must include all emissions.[117] A single permit may be issued for a facility with multiple sources.[118] Permits must be submitted by the states to the EPA for review;[119] they are enforceable not only by state authorities but also by federal authorities.[120] The EPA published its federal operating permit program rules on July 21, 1992.[121] States were directed to submit their programs by November 15, 1993 for approval or disapproval by EPA within one year or November 15, 1994.

How the Programs Work Together

The above programs can be viewed as follows:

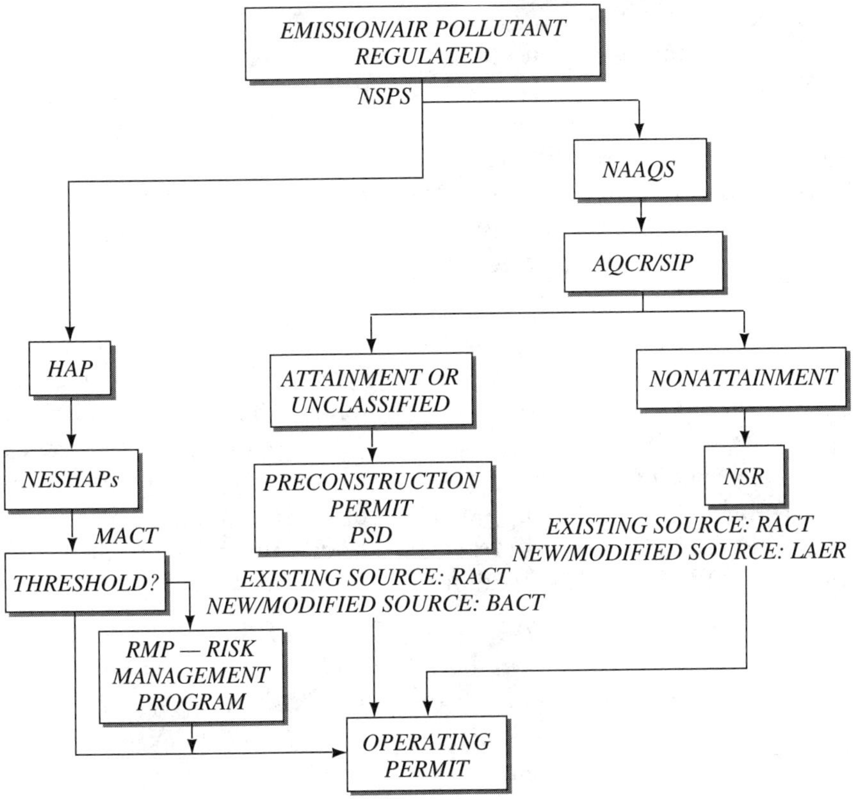

Enforcement and Penalties

The CAA provides for enforcement by civil action by the EPA, administrative action by the EPA, criminal actions, and citizen suits.

Civil

The EPA can issue a compliance order to a person who has violated or is violating any requirement of the CAA.[122] Civil penalties are provided for violation of a compliance order.[123] The EPA can also bring an action in federal district court seeking injunctive relief or civil penalties of up to $25,000 per day for current or past violations of CAA requirements.[124]

Administrative

The EPA has broad authority to impose administrative penalties without going to court.[125] An administrative order can be issued to assess penalties of up to $25,000 per day of violation of many of the CAA requirements (including permit requirements)[126] up to a maximum of $200,000.[127] An administrative hearing is required for these penalties,[128] but "field citations" of up to $5,000 per day of violation are also authorized.[129] The field citation is like a traffic citation—the person cited can either pay the specified penalty or request a hearing.[130] The issuance and payment of a field citation do not preclude other enforcement actions.[131] A "noncompliance" penalty can be assessed, which equals the economic benefit the source received by not complying with CAA requirements.[132]

Criminal

Criminal penalties are authorized for several "knowing" violations of the CAA.[133] Penalties can be as much as $500,000 for an organization, and individuals can face prison terms of up to five years for the first violation.[134] Further, a person who "negligently" or "knowingly" releases specified HAPs and places another person in "imminent danger of death or serious bodily injury" faces criminal penalties.[135] "Knowingly" is defined:

> *In determining whether a defendant who is an individual knew that the violation placed another person in imminent danger of death or serious bodily injury—*
>
> *(i) the defendant is responsible only for actual awareness or actual belief possessed; and*

(ii) *knowledge possessed by a person other than the defendant, but not by the defendant, may not be attributed to the defendant; except that in proving a defendant's possession of actual knowledge, circumstantial evidence may be used, including evidence that the defendant took affirmative steps to be shielded from relevant information.*[136]

Citizen Suits

Citizen suits in federal district court are authorized against persons alleged to be in violation of an emission standard or limitation, or an order issued by the EPA or a state.[137] An "emission standard or limitation" is defined broadly and includes requirements of permits.[138] Suit may also be brought against persons operating without a permit or in violation of a permit issued under PSD or nonattainment programs.[139] A citizen must provide notice to the EPA, the state, and the alleged violator at least 60 days prior to filing suit.[140]

Bounty

The EPA Administrator may pay an award of up to $10,000 to any person who furnishes information or services leading to a criminal conviction or a civil penalty (judicial or administrative).[141]

Conclusion

As the CAA has evolved, it has numerous requirements that apply to E&P. The provisions encompass many activities and impose requirements for preconstruction and operating permits. An E&P operator must identify emissions from the facilities by type and amount to determine what provisions apply. Permits must be obtained and their provisions followed. The operator must also ensure compliance with the requirements of the SIP.

APPENDIX 9 ENDNOTES

1. Clean Air Act Amendments of 1990, Pub L. 101-549, 104 Stat. 2399; Air Pollution Control Act of 1955, Pub.L. 81-145, 69 Stat. 322; Steel Industry Compliance Extension Act of 1981, Pub.L. 97.23, 95 Stat. 139; Clean Air Act of 1963, Pub.L. 88-206, 77 Stat. 392; Motor Vehicle Air Pollution Control Act, Pub.L. No. 89.272, 79 Stat. 992; Clean Air Act Amendments of 1966, Pub.L. No. 89-675, 80 Stat. 954; Air Quality Act of 1967, Pub. L.

No. 90-148, 81 Stat. 485; Clean Air Act Amendments of 1970, Pub.L. No. 91-604, 84 Stat. 1676; Comprehensive Health Manpower Training Act of 1971, Pub.L. 92-157, 85 Stat. 431; Energy Supply and Environmental Coordination Act of 1974, Pub.L. No. 93-319, 88 Stat. 246; Clean Air Act Amendments of 1977, Pub.L. No. 95-95, 91 Stat. 686. William H. Rodgers, Jr., *Environmental Law* 124, N.1 (2d ed. 1994).

2. 42 USC § 7401(b)(1).
3. 42 USC § 7401(b)(2).
4. 42 USC § 7401(b)(3).
5. 42 USC § 7401(b)(4).
6. 42 USC § 7401(c).
7. Z.J.B. Plater, et al., *Environmental Law and Policy: Nature, Law and Society,* 772-73 (1992).
8. The techniques and structure could also be approached by the Titles of the Clean Air Act Amendments of 1990: Title I—Provisions for Attainment and Maintenance of National Ambient Air Quality Standards; Title II—Provisions Relating to Mobile Sources; Title III—Hazardous Air Pollutants; Title IV—Acid Deposition Control; Title V—Permits; Title VI—Stratospheric Ozone Protection; Title VII—Provisions Relating to Enforcement: Title VIII—Miscellaneous Provisions. Pub.L. 101-549.
9. 42 USC § 7408.
10. "The term "air pollutant" means any air pollution agent or combination of such agents, including any physical, chemical, biological, radioactive (including source material, special nuclear material, and byproduct material) substance or matter which is emitted into or otherwise enters the ambient air. Such term includes any precursors to the formation of any air pollutant, to the extent the Administrator has identified such precursor or precursors for the particular purpose for which the term "air pollutant" is used." 42 USC § 7602(g).
11. 42 USC § 7480(a)(1)(A-B).
12. 42 USC § 7408(a)(2).
13. 42 USC § 7408(b)(1).
14. 42 USC § 7408(d).
15. 42 USC § 7409(a)(2).
16. 42 USC § 7602(u).
17. 42 USC § 7409(b)(1).
18. 42 USC § 7409(b)(2).
19. ". . . effects on welfare includes, but is not limited to, effects on soils, water, crops, vegetation, manmade materials, animals, wildlife, weather, visibility, and climate, damage to and deterioration of property, and hazards to transportation, as well as effects on economic values and on personal comfort and well-being, whether caused by transformation, conversion, or combination with other air pollutants." 42 USC § 7602(h).
20. 42 USC § 7602(g).
21. 42 USC § 7410(a)(1).
22. 42 USC § 7410(a)(2).
23. 42 USC §§ 7410(a)(2)(A)-(M).
24. 42 USC Part C.

25. 42 USC § 7471.
26. 42 USC § 7502(c).
27. 42 USC § 7412(1).
28. 42 USC § 7407(a).
29. 42 USC § 7407.
30. 42 USC § 7407(b).
31. 42 USC § 7407(a).
32. 42 USC § 7407(d)(1)(A).
33. 42 USC § 7502(a)(2)(A).
34. 42 USC § 7471.
35. 42 USC § 7401(b)(1), 7460-7469a; Sierra Club v. Ruckelshaus, 344 F. Supp. 253 (D.D.C. 1972).
36. 42 USC § 7602(g).
37. 42 USC § 7479(1).
38. 42 USC § 7472.
39. 42 USC § 7473(b)(1).
40. 42 USC § 7473(b)(2).
41. 42 USC § 7473(b)(3).
42. 42 USC § 7476.
43. 42 USC § 7473(b)(4).
44. 42 USC § 7475(a)(1)-(2).
45. 42 USC § 7475(a)(4).
46. 42 USC § 7491.
47. 42 USC § 7502(a)(2)(A).
48. 42 USC § 7471.
49. 42 USC § 7502(b)(5).
50. 42 USC § 7502(c)(1)-(9).
51. 42 USC § 7503(a)(1)(A).
52. 42 USC § 7503(a)(1)(B).
53. 42 USC § 7503(a)(2).
54. 42 USC § 7503(a)(3).
55. 42 USC § 7503(a)(5).
56. 42 USC § 7503(c)(1).
57. 42 USC § 7509(a).
58. 42 USC § 7509(b)(1)-(2).
59. 42 USC § 7602(j).
60. 42 USC § 7511a(a).
61. 42 USC § 7511a(a)(4).
62. 42 USC § 7511a(b).
63. 42 USC § 7511a(b)(5).
64. 42 USC § 7511a(c).
65. 42 USC § 7511a(c)(10).
66. 42 USC § 7511a(d).
67. 42 USC § 7511a(d)(2).
68. 42 USC § 7511a(e).
69. 42 USC § 7511a(e)(1).
70. 42 USC § 7472(b).
71. 42 USC § 7411.

72. 42 USC § 7411(a)(1).
73. 42 USC § 7411(b)(1)(B).
74. 42 USC § 7411(a)(2).
75. 42 USC § 7411(a)(3).
76. 42 USC § 7411(b)(1)(A).
77. 42 USC § 7411(b)(2)(B).
78. 42 USC § 7411(d)(1); 42 USC § 7411(c)(1).
79. 42 USC § 7411(e).
80. 42 USC § 7411(h)(1).
81. 40 CFR Part 60.
82. 42 USC § 7412.
83. 42 USC § 7412(a)(6).
84. 42 USC § 7412(b)(1).
85. 42 USC § 7412(b)(2).
86. 42 USC § 7412(a)(1).
87. 42 USC § 7412(a)(2).
88. 42 USC § 7412(c)(1).
89. 42 USC § 7412(c)(2).
90. 42 USC § 7412(d)(1).
91. 42 USC § 7412(d)(2).
92. 42 USC § 7412(d)(2)(A)-(E).
93. 42 USC § 7412(d)(3).
94. 42 USC § 7412(d)(7).
95. 42 USC § 7412(d)(3).
96. 42 USC § 7412(d)(3)(A)-(B).
97. 42 USC § 7412(g)(2)(A)-(B).
98. 42 USC § 7412(b)(6).
99. 42 USC § 7412(r)(1).
100. 42 USC § 7412(r)(1).
101. 42 USC § 7412(r)(4).
102. 42 USC § 7412(r)(5).
103. Proposed Rules, 58 Fed. Reg. 5102 (1990) (to be codified at 40 CFR Part 68).
104. Proposed Rules, 58 Fed. Reg. 54,190, 54,191 (1990) (to be codified at 40 CFR Part 68).
105. Rules and Regulations, 59 Fed. Reg. 4378 (1994) (to be codified at 29 CFR Part 1910).
106. 42 USC § 7412(r)(7)(B)(i)-(ii).
107. 42 USC § 7412(r)(7)(B)(ii)(I)-(III).
108. 42 USC § 7412(r)(7)(B)(iii).
109. 42 USC § 7408(h).
110. 42 USC § 7479(3).
111. 42 USC § 7501(3).
112. 42 USC § 7412(d)(2).
113. 42 USC § 7412(d)(3).
114. 42 USC § 7411(a)(2).
115. 42 USC § § 7661-7661f.
116. 42 USC § 7661a(a).

117. 42 USC § 7661a(a).
118. 42 USC § 7661a(c).
119. 42 USC § 7661d.
120. 42 USC § 7602(k).
121. 40 CFR 70.
122. 42 USC § 7413(a).
123. 42 USC §§ 7413(a)(1), (3) and 7413(b).
124. 42 USC § 7413(b).
125. 42 USC § 7413(d).
126. 42 USC § 7413(1)(A)-(C).
127. 42 USC § 7413(d)(1).
128. 42 USC § 7413(d)(2)(A).
129. 42 USC § 7413(d)(3).
130. 42 USC § 7413(d)(3).
131. 42 USC § 7413(d)(3).
132. 42 USC § 7420.
133. 42 USC § 7413(c).
134. 42 USC § 7413(c)(1)-(3).
135. 42 USC § 7413(c)(3)-(4).
136. 42 USC § 7413(c)(5)(B).
137. 42 USC § 7604(a)(1).
138. 42 USC § 7604(f).
139. 42 USC § 7604(a)(3).
140. 42 USC § 7604(b).
141. 42 USC § 7413(f).

APPENDIX 10
Endangered Species Act Detailed Summary

As the human population of the United States expands, two trends can be observed. First, the expanding population is moving into previously less developed areas of the country, resulting in a reduction of the undeveloped habitat available for wildlife. This trend has resulted in a corresponding increase in the number of species that are listed as threatened or endangered under the provisions of the Endangered Species Act of 1973 (ESA).[1] Second, the population's increasing demand for energy has led exploration and production operations to expand into more remote, pristine areas. The effect of these coincident trends has been an increase in the number of cases in which the needs of listed species must be balanced against the need to identify and produce new oil and gas resources.

In order to understand the potential magnitude of the impacts of the ESA on exploration and production projects, it is first necessary to understand the magnitude of the problem the act was intended to address. Species are currently becoming extinct at an unprecedented rate. Within the scientific community, it is generally accepted that we are losing species at such a rate that many, particularly in little studied and rapidly disappearing rainforest habitats, will be extinct before they are scientifically described. Estimates of the total number of species on earth vary from 3 million to 30 million. Of these, only about 1.4 million have been formally named and described by science. Of the species thus far described, approximately 700,000 are insects; 125,000 are arthropods; 250,000 are flowering plants; 70,000 are fungi; and 45,000 are vertebrates. Algae and microorganisms account for the remaining 100,000.

The rate of extinction of species is also subject to debate, but the range of variation for current estimates is considerably smaller. Professor Edward O. Wilson of Harvard University places the current rate of species extinction at approximately 27,000 per year in tropical rainforests alone. Dr. Peter Raven of the Missouri Botanical Gardens places the estimated extinction rate at approximately 100 per day, or 36,500 per year for all ecosystems on earth combined. At the present rate of habitat destruction, he estimates that this rate of extinction could be maintained for three decades, culminating in the loss of over one million species. Although extinction is acknowledged to be part of evolution, this rate is conservatively estimated to be at least 1,000 times the prehistoric rate. As Alastair S. Gunn points out,

> *"The wholesale slaughter of entire species by human actions bears no resemblance to the evolutionary process of speciation: evolution and the theory of natural selection can give an account of the passing of the dinosaurs, but not of the extermination of the passenger pigeon".*[2]

It is this situation that was specifically acknowledged by Congress in its statement of intent for the ESA. "Congress finds and declares that

(1) various species of fish, wildlife, and plants in the United States have been rendered extinct as a consequence of economic growth and development untempered by adequate concern and conservation;

(2) other species of fish, wildlife and plants have been so depleted in numbers that they are in danger of or threatened with extinction;

(3) these species of fish, wildlife and plants are of aesthetic, ecological, educational, historical, recreational, and of scientific value to the Nation and to its people".[3]

Species listed under the ESA are found in all 50 states and the territorial seas of the United States (Figure 10–1). Table 10–1 gives the total numbers of listed species in the United States by taxonomic group as of January 1, 1995.

As these figures demonstrate, it is possible to encounter listed species anywhere in the country while exploring for or producing oil and gas.

Purpose of the ESA

Congress' purpose for the ESA is to "provide a means whereby the ecosystems upon which endangered species depend may be conserved", and "to provide a program for the conservation of such endangered and threatened species".[5] The stated policy, intended to further this purpose, is that "all Federal departments

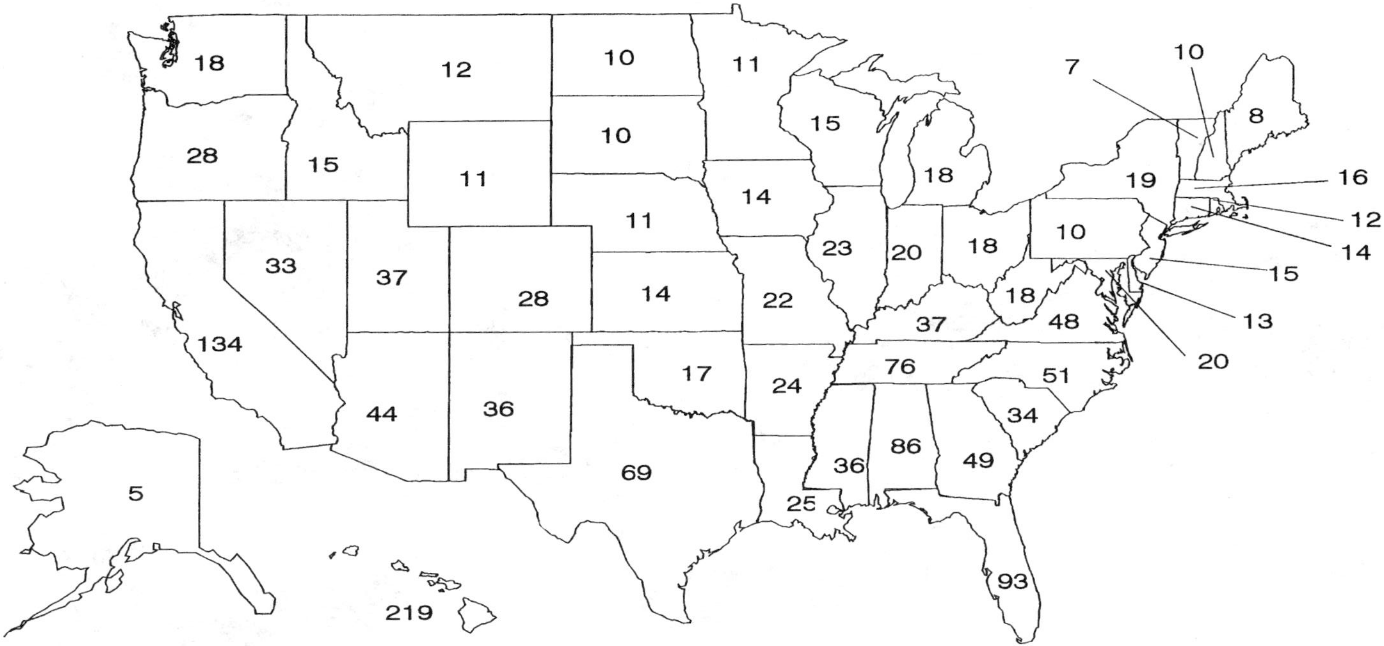

Figure 10–1 United States Map with Listed Species Number. Source: U.S. Fish and Wildlife Service (1994).

Table 10–1 Listed Species in the United States[4]

Group	Endangered	Threatened	Total
Mammals	55	9	64
Birds	75	16	91
Reptiles	14	19	33
Amphibians	7	5	12
Fishes	68	37	105
Snails	15	7	22
Clams	51	6	57
Crustaceans	14	3	17
Insects	19	9	28
Arachnids	4	0	4
ANIMAL SUBTOTAL	322	111	433
Plants	422	89	511
GRAND TOTAL	744	200	944

and agencies shall seek to conserve endangered species and threatened species and shall utilize their authorities in the furtherance of the purposes of this chapter".[6] The statutory and regulatory provisions that are discussed in subsequent portions of this chapter are intended to accomplish this purpose.

Key Provisions of the ESA

The sections of the ESA that are most likely to effect exploration and production operations are the following:

- Section 4, Determination of Endangered Species and Threatened Species;[7]
- Section 7, Interagency Cooperation;[8]
- Section 9, Prohibited Acts;[9] and
- Section 10, Exceptions.[10]

Each of these sections is discussed in detail below.

ESA *Section 4—Determination of* Endangered *Species and Threatened Species*

The key to species protection under the ESA is its listing as either endangered or threatened by the Secretary of the Interior under the provisions of Section 4. Only those species that have been listed under this section receive the protection of the act.

Under the ESA, the term "species" includes "any subspecies of fish, or wildlife or plants, and any distinct population of any species of vertebrate fish or wildlife which interbreeds when mature."[11] This definition is considerably broader than the biological definition of species, which is typically given as any population of organisms, potentially interbreeding, that are reproductively isolated from other populations, and produce viable offspring. This statutory definition of a species allows the Secretary to extend the protection of the Endangered Species Act to segments of biological species, with or without any actual taxonomic distinction from other populations.

The Endangered Species Act defines an endangered species as "any species which is in danger of extinction throughout all or a significant portion of its range."[12] A threatened species is defined as "any species which is likely to become an endangered species within the foreseeable future throughout all or a significant portion of its range."[13] Both of these definitions exclude members of "the Class Insecta determined by the Secretary to constitute a pest whose protection under the provisions of this chapter would present an overwhelming and overriding risk to man,"[14] thus avoiding the potential political fallout from a proposed listing of, for example, the *Culex* mosquito.

The combination of these definitions has allowed for the listing of complete species, subspecies, and regional populations of species with no subspecies status. Recent lists that demonstrate the scope of these definitions include the following:

- The golden-cheeked warbler (*Dendroica chrysoparia*) was listed as endangered because it was determined that the species was in danger of extinction throughout its entire range. This is the more traditionally acknowledged form of species endangerment.
- The black-capped vireo (*Vireo atricapillus*) was listed as endangered because the species was considered to be in danger of extinction in a significant portion of its range. The species as a whole appears to be in no imminent danger of extinction, since Mexican populations appear sound. However, the entire species was listed for two reasons: since the population in the area of endangerment is not

taxonomically distinct from the remainder of the population, and since the species is highly mobile, and interchange between populations is at least theoretically possible.

- The Texas trailing phlox (*Phlox nivalis* ssp. *texensis*) was listed as endangered because it is a subspecies considered to be in danger of extinction throughout its entire range. The species as a whole is not listed.
- The gopher tortoise (*Gopherus polyphemus*) was partially listed as threatened. This listing occurred because the population located west of the Mobile and Tombigbee Rivers, in Alabama, Mississippi, and Louisiana, is considered likely to become endangered in the foreseeable future. The area west of these rivers is considered to be a "significant portion of the range" of this species. Those individuals living in this area are considered to constitute a "distinct population segment." It is the combination of these factors that allows for the partial listing. Populations of the gopher tortoise located east of the Mobile and Tombigbee Rivers are not listed.

Administrative Procedures for Listing. When evaluating a species for listing, the Secretary of the Interior is to consider five factors:

- The present or threatened destruction, modification, or curtailment of the species habitat or range;
- Over-utilization for commercial, recreational, scientific, or educational purposes;
- Disease or predation;
- The inadequacy of existing regulatory mechanisms;
- Other factors affecting the species continued existence.[15]

The determination of these factors is to be made "solely on the basis of the best scientific and commercial information . . . without reference to possible economic or other impacts of such determinations."[16] That it is the intent of Congress that economic impacts of a listing not be considered is reinforced by reports from the debates of the 1982 amendments to the ESA in which it was stated "economic considerations have no relevance to determinations of species status."[17]

For a normal listing, there are two administrative steps that may precede a species being listed as threatened or endangered. It may first be treated as a candidate species and then as a proposed species.

A candidate species is one for which the Secretary has received a petition from an interested party requesting that a rule be promulgated to list the species in question.[18] The candidate status remains in effect during the period in which the Secretary conducts a review of the petition to determine if it is substantially supported by "scientific or commercial information indicating that the petitioned action may be warranted."[19] Although a candidate species is not actually protected under the ESA, it is generally given consideration in decision making, since accelerated procedures exist that could result in the promulgation of a rule listing the species before the proposed project is completed.[20] Generally, the Secretary has 12 months from the receipt of the petition to publish a finding that the petitioned action is not warranted,[21] that it is warranted,[22] or that it is warranted but other pending proposals preclude timely action on it.[23] It should be noted that petitions determined to have insufficient data to support the petitioned action can be revised and resubmitted. This allows the Secretary an additional 12 months to review and act on the resubmitted petition.[24]

Two candidate designations are commonly seen in correspondence with the U.S. Fish and Wildlife Service (FWS) and the National Marine Fisheries Service (NMFS):

- C1—A candidate for federal listing, with enough substantial information on biological vulnerability and threats to support proposals for listing; and
- C2—A candidate for federal listing with some evidence of vulnerability, but for which not enough data exists to support listing.

If, upon review of the best scientific and commercial data available, the Secretary then determines that a proposed rule to list the species as threatened or endangered should be published in the Federal Register, the species will become a proposed species.

A proposed species is one for which an actual notice of proposed rule making has been published in the Federal Register.[25] The proposal can be to list the species as either threatened or endangered. As in the case of candidate status, proposed status does not convey the protection of the ESA to a species. But when contemplating an action that could affect a proposed species, it is good policy to take such species into account since the proposed rule may be final before the completion of the project being considered.

The Secretary is required to allow at least 60 days for public comment following the publication in the Federal Register of a proposed rule to list a species.[26] This time period may be (and with listing proposals frequently is) extended or reopened if the Secretary finds there is good cause to do so. If any person so requests, the Secretary is required to hold at least one public hearing within 45 days of publication of the proposed rule.

Within one year of the publication of the proposed rule, the Secretary must either:

- Publish a final rule implementing the proposed rule;
- Publish a notice withdrawing the proposed rule; or
- Publish a notice extending the review period by up to six months because there is substantial disagreement among scientists knowledgeable about the species concerned regarding the sufficiency or accuracy of available data relevant to the proposed listing.[27]

Emergency Listings. The provisions for emergency listings of species provide that the Secretary may publish a rule in the Federal Register listing a publication. To follow this procedure, the Secretary must also give notice to the relevant state agency in each state where the listed species occurs. An emergency listing is effective for 240 days, during which time the normal rule-making procedures must be complied with in order for the rule to remain in effect. The emergency rule-making procedures may be invoked by the Secretary in any situation posing a significant risk to the well-being of any species of fish, wildlife, or plant.[28]

Designation of Critical Habitat. Section 4 of the ESA also provides for the designation of critical habitat for listed species. Critical habitat may be designated for a threatened or endangered species either at the time of listing or subsequent to it, if later information indicates that such action is warranted. Critical habitat is defined under the act as:

> *(i) the specific areas within the geographical areas occupied by the species, at the time it is listed in accordance*

> *with the provisions of section 1533 of this title, on which are found those physical or biological features (I) essential to the conservation of the species and (II) which may require special management consideration or protection and (ii) specific areas outside the geographical area occupied by the species at the time it is listed in accordance with the provisions of section 1533 of this title, upon a determination by the Secretary that such areas are essential for the conservation of the species.*[29]

Procedure for the designation of critical habitat is similar to that for the listing of species.

Section 4 Impacts on Exploration and Production Projects. Since the status of species listings and critical habitat designation can have major effects on exploration and production activities, these provisions of Section 4 of the ESA are critical to project planning. By monitoring the status of candidate species and proposed listings, project planners can determine the degree to which project modification may be required to accommodate listed species. Emergency listings are a perpetual hazard for project planners. Thorough knowledge of species status in a project area can help avoid problems arising from emergency listings.

An example of what can happen as a result of emergency listings where species information is not adequate, is the Kern River pipeline project in California. An emergency listing was granted for the desert tortoise (*Gopherus agassizii*). A Biological Opinion issued by the FWS contained a takings clause allowing incidental takings of 25 tortoises. Due to unexpectedly high populations, the takings were made within the first two weeks of construction. Takings beyond those specifically authorized were subject to fines of up to $50,000 per individual tortoise taken. As a result of this, it was necessary to employ biologists to continuously monitor the right of way. All contractor personnel were trained to recognize and protect the tortoises and speeds on area roadways were limited to 15 miles per hour to avoid hitting tortoises.[30]

Critical habitat designated under Section 4 of the ESA can also affect exploration and production projects. Since many of these projects require federal permits, the section 7 requirement that federal agencies "shall in consultation with and with the assistance of the Secretary, insure that any action authorized, funded or carried out by such agency is not likely to . . . result in the destruction or adverse modification of "designated critical habitat."[31] This requirement can result in the necessity to modify projects to avoid unacceptable impacts, or in extreme cases, cancel the project.

ESA *Section 7—Interagency Cooperation*

Interagency cooperation under Section 7 of the ESA is the most likely route for the act to make an impact on exploration and production activities. These provisions can be encountered any time exploration and production activities require a federal permit or are undertaken on federal property. The process has been developed to implement the congressional mandate that all federal agencies shall seek to conserve endangered and threatened species. The implementation of Section 7 requires that all federal agencies consult with FWS and/or NMFS in order to insure that their actions do not jeopardize the continued existence of listed species. Figure 10–2 shows a diagrammatic representation of the Section 7 consultation process. This process has the advantage of having reasonably predictable statutory and regulatory time lines. It can also, in some cases, eliminate the need for takings permits under Section 10 of the ESA through the use of taking statements in the Biological Opinion.

The Consultation Process. The consultation process may be carried out by representatives of the lead federal agency or by a designated non-federal representative.[32] In most exploration and production cases, the non-federal representative will be the applicant for the lease or permit. The federal agency in such cases is supposed to provide guidance to the non-federal representative. The ultimate responsibility for Section 7 compliance, however, remains with the federal agency.

Initiating Consultation. Section 7 consultation is ordinarily initiated with a request for information by the federal agency or non-federal representative to FWS and/or NMFS (the service). Typically, these requests are submitted to the Ecological Services Office of FWS with responsibility for the geographic area of the project. The request must be in writing and should ask for a list of any listed or proposed species, and any designated or proposed critical habitat in the project area.[33] The response, which is typically prepared by designated Section 7 coordinators in the regional office, must be sent within 30 days of the receipt of the

Figure 10-2 Section 7 Consultation Process for Federal Agency Action

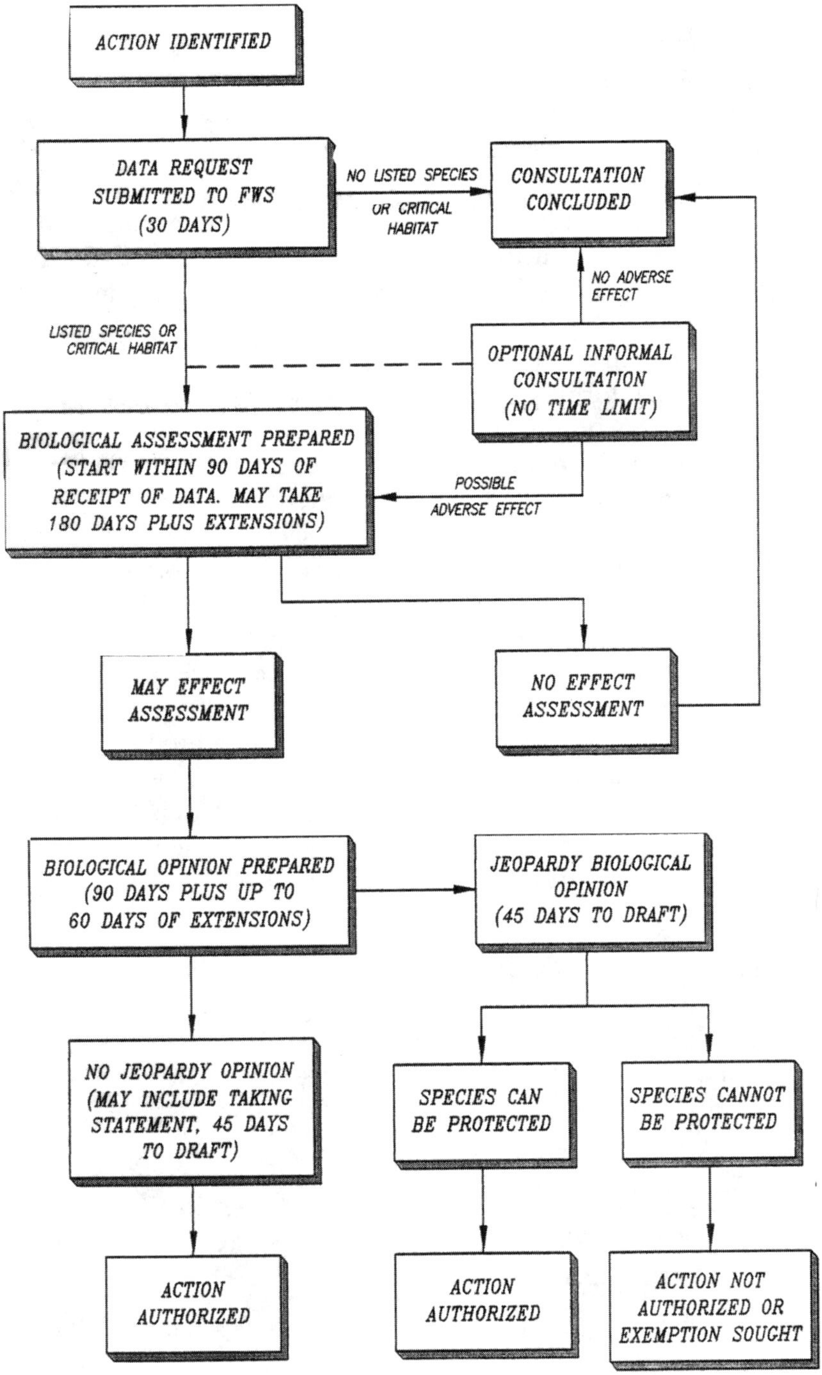

request. If the service advises that no listed or proposed species, or designated or proposed critical habitat are present in the project's impact area, the lead agency need not prepare a biological assessment nor is further consultation required.[34]

Informal Consultation. Informal Consultation is an optional process provided for by regulation. It consists of all discussions and correspondence between the service and the lead agency or non-federal representative. The process is intended to provide guidance in determining if the implementation of formal consultation is necessary. If during informal consultation it is determined that the action is unlikely to adversely affect listed species or critical habitat, the consultation process is concluded and no further action is necessary.[35]

During the informal consultation, the service may suggest modifications to the proposed action that could be implemented to avoid the likelihood of adverse impacts to listed species or critical habitats.[36] These modifications could include conducting the project in stages to avoid disruption to excessive areas of habitat at any given time, working only at certain times of the day or year to avoid interfering with essential species behavior patterns, or otherwise implementing protective measures to avoid incidental takings.

Informal consultation should be regarded as a potentially invaluable tool to move projects forward without the burden of formal consultation. It can be highly successful when entered into with an open-minded approach and willingness to make reasonable compromises where the project will allow. Skillfully conducted, informal consultation can lay an excellent foundation for cooperation between project applicants, federal lead agencies, and the service.

Biological Assessment. In order for the lead agency to determine if formal consultation with the service is required for a particular federal action, the lead agency or its designated non-federal representative must prepare a Biological Assessment.[37] Although the contents of the Biological Assessment are at the discretion of the lead agency, regulations state that the following may be considered for inclusion:

(1) The results of an on-site inspection of the area affected by the action to determine if listed or proposed species are present or occur seasonally.

(2) The views of recognized experts on the species at issue.

(3) A review of literature and other information.

(4) An analysis of the effects of the action on the species and habitat, including considerations of cumulative effects, and the results of any related studies.

(5) An analysis of alternate actions by the Federal agency for the proposed action.[38]

In order to assure that the Biological Assessment addresses current concerns, it must be commenced within 90 days of the receipt of the information requested from the Service.[39] The lead agency is required to complete the Biological Assessment within 180 days unless some other time period has been arranged. If a permit or license applicant is involved, the time may not be extended beyond 180 days without their concurrence.[40] Since the improvement of data quality is the normal objective in extending this time period, it will often be in the applicant's best interest to agree to reasonable extensions with the intent of ultimately expediting the process.

The purpose of the Biological Assessment is to aid the lead agency in determining if the proposed action is likely to adversely effect listed species or critical habitat.[41] Based on the results of the Biological Assessment, the lead agency will make either a "no effect" or a "may effect" determination as to the proposed action. The Biological Assessment is then submitted to the service for concurrence. If there has been a "no effect" determination and the service concurs, then no further consultation is required. On the other hand, if a "may effect" determination is made and the service concurs, formal consultation will be initiated. The service has 30 days from the receipt of the Biological Assessment to respond. The lead agency may then choose to request the initiation of formal consultation at the time of submittal of the Biological Assessment, in order to expedite the process.[42]

Formal Consultation. Formal consultation is to be initiated at the earliest point the lead agency determines that the proposed action may affect listed species or critical habitats.[43] Formal consultation must be initiated by a written request from the lead agency. This request shall include:

(1) A description of the action to be considered;

(2) A description of the specific area that may be affected by the action;

(3) A description of any listed species or critical habitat that may be affected by the action;

(4) A description of the manner in which the action may affect any listed species or critical habitat and an analysis of any cumulative effects;

(5) Relevant reports, including any environmental impact statement, environmental assessment, or biological assessment that may have been prepared; and

(6) Any other relevant available information on the action, the affected listed species, or critical habitat.[44]

The lead agency is responsible for providing the service with the best scientific and commercial data available.[45] The service is responsible for the review of this data. The formal consultation must conclude within 90 days after its initiation unless it is extended. Where an applicant for a license or permit is involved, the service must submit to the applicant, "a written statement setting forth:

(1) The reasons why a longer time is required,

(2) The information that is required to complete the consultation, and

(3) The estimated date on which the consultation will be completed."[46]

Consultation involving an applicant cannot be extended for more than 60 days without his consent. Within 45 days of conclusion of the formal consultation, the service is supposed to deliver a Biological Opinion to the lead agency and the applicant.[47]

The Biological Opinion is the opinion of the service as to whether or not the proposed action will jeopardize the continued existence of any listed species or critical habitat. Biological Opinions are generally referred to as either "no jeopardy" or "jeopardy" opinions.

When the service issues a "no jeopardy biological opinion," it may include an incidental taking statement. The take must be incidental to an otherwise lawful action. The statement must be such that it:

(i) Specifies the impact, i.e., the amount or extent, of such incidental taking on the species;

(ii) Specifies those reasonable and prudent measures that the Service considers necessary or appropriate to minimize such impact;

(iii) In the case of marine mammals, specifies those measures that are necessary to comply with Section 101 (a) (5) of the Marine Mammal Protection Act of 1972 and applicable regulations with regard to such taking;

(iv) Sets forth the terms and conditions (including, but not limited to reporting requirements) that must be complied with by the Federal Agency or any applicant to implement the measures specified under paragraphs (i) (1) (iii) of this section; and

(v) Specifies the procedure to be used to handle or dispose of any individuals of a species actually taken.[48]

It should be noted that "reasonable and prudent measures, along with the terms and conditions that implement them, cannot alter the basic design, location, scope, duration or timing of the action and may invoke minor changes only."[49]

Formal consultation is concluded with the issuance of the Biological Opinion. While the time consumed by the process can be over a year by the time all possible extensions and reporting periods are maximized, this process does have a reasonably predictable timetable when compared with takings permits for private action under Section 10 of the ESA (discussed in the next section). The incorporation of an incidental taking statement in a biological opinion is also a distinct advantage of the Section 7 consultation process, since it eliminates the need to obtain permits through the Section 10 procedure. It should be observed, however, that taking statements are strictly enforced. Take in excess of that specified in the incidental takings statement could potentially subject the responsible parties to the criminal provisions of the act.

In those instances where a jeopardy biological opinion is issued, the lead agency must notify the service as to its final decision on the action.[50] If the agency determines that it cannot comply with its obligation to insure that its action does not jeopardize the continued existence of any listed species, it can either not take the action or apply for an exemption. The exemption process provided for federal action is almost never utilized and will not be discussed here.

ESA Section 9—Prohibited Acts

Prohibited Acts under the ESA are described under Section 9. The principal prohibited act under the ESA with which exploration and production projects are likely to be concerned is "taking." Taking is broadly defined under the act and can include such actions as "harass, harm, pursue, hurt, shoot, wound, kill, trap, capture, or collect, or attempt to engage in such conduct."[51] The regulations expand the definitions of harass and harm in such a way that the actions regulated are expanded beyond the apparent meaning of the act. "Harass" in the definition of "take" in the Act means "an intentional or negligent act or omission which creates the likelihood of injury to wildlife by annoying it to such an extent as to significantly disrupt normal behavior patterns which include but are not limited to breeding, feeding or sheltering."[52] "Harm" in the definition of "take" in the Act means "an act which actually kills or injures wildlife. Such acts may include significant habitat modification or degradation where it actually kills or injures wildlife by significantly impairing essential behavioral patterns including breeding, feeding or sheltering."[53] As will be discussed, this regulatory definition of harm has been the focus of much of the recent ESA controversy. "Take" need not entail any intentional act. Unintentional taking is referred to as "incidental taking" and is defined as: "any taking otherwise prohibited, if such taking is incidental to, and not the purpose of, the carrying out of an otherwise lawful activity."[54]

Takings under Section 9 are prohibited in purely private actions as well as federal agency actions. As such, the prohibition on takings can reach aspects of exploration and production operations that require no federal authorization.[55]

ESA Section 10—Incidental Takings Permits for Private Actions

Section 10 of the ESA actually deals with all exceptions to the provisions of the act. The exception that is likely to be of the greatest importance for exploration and production activities, however, are the incidental takings permits provided for by Section 10 (a) (1) (B) of the ESA.[56]

While such permits are available under the act, their practical significance is somewhat reduced due to the high burden placed on the applicant. The ESA provides that the Secretary shall not authorize any taking under the act "unless the applicant therefore submits to the Secretary a conservation plan that specifies:

(i) the impact that will likely result from such taking;

(ii) what steps the applicant will take to minimize and mitigate such impacts, and the funding that will be available to implement such steps;

(iii) what alternative actions to such takings the applicant considered and the reasons why such alternatives are not being utilized; and

(iv) such other measures that Secretary may require as being necessary or appropriate for purposes of the plan."[57]

The Secretary is to review the plan, provide an opportunity for public comment and, it is found that:

(i) the taking will be incidental;

(ii) the applicant will, to the maximum extent practicable, minimize and mitigate the impacts of such taking;

(iii) the applicant will ensure that adequate funding for the plan will be provided;

(iv) the taking will not appreciably reduce the likelihood of the survival and recovery of the species in the wild; and

(v) the measures, if any, under (A) (iv) will be met.[58]

The Secretary shall issue the permit. The permit may be revoked if, at any time, it is found that the applicant is not complying with the terms of the permit.[59]

The regulations under this section require that application for such taking permits be made on an official application form and that the applicant specify common and scientific names of the species to be taken, as well as the number, age, and sex of such species, if known.[60]

Very few incidental taking permits are issued for purely private actions. It was the intent of Congress that permits not be issued unless the applicant is prepared to make substantial concessions for species preservation. There is no mandated time line for such permits. Applicants frequently find that the expense of employing consultants and attorneys to gather large amounts of biological data and negotiate permit conditions make the project prohibitively costly, and they either drop the project or choose an alternative. The prospect of a long-term commitment of resources for

Figure 10–3 Incidental Taking Permits for Private Actions

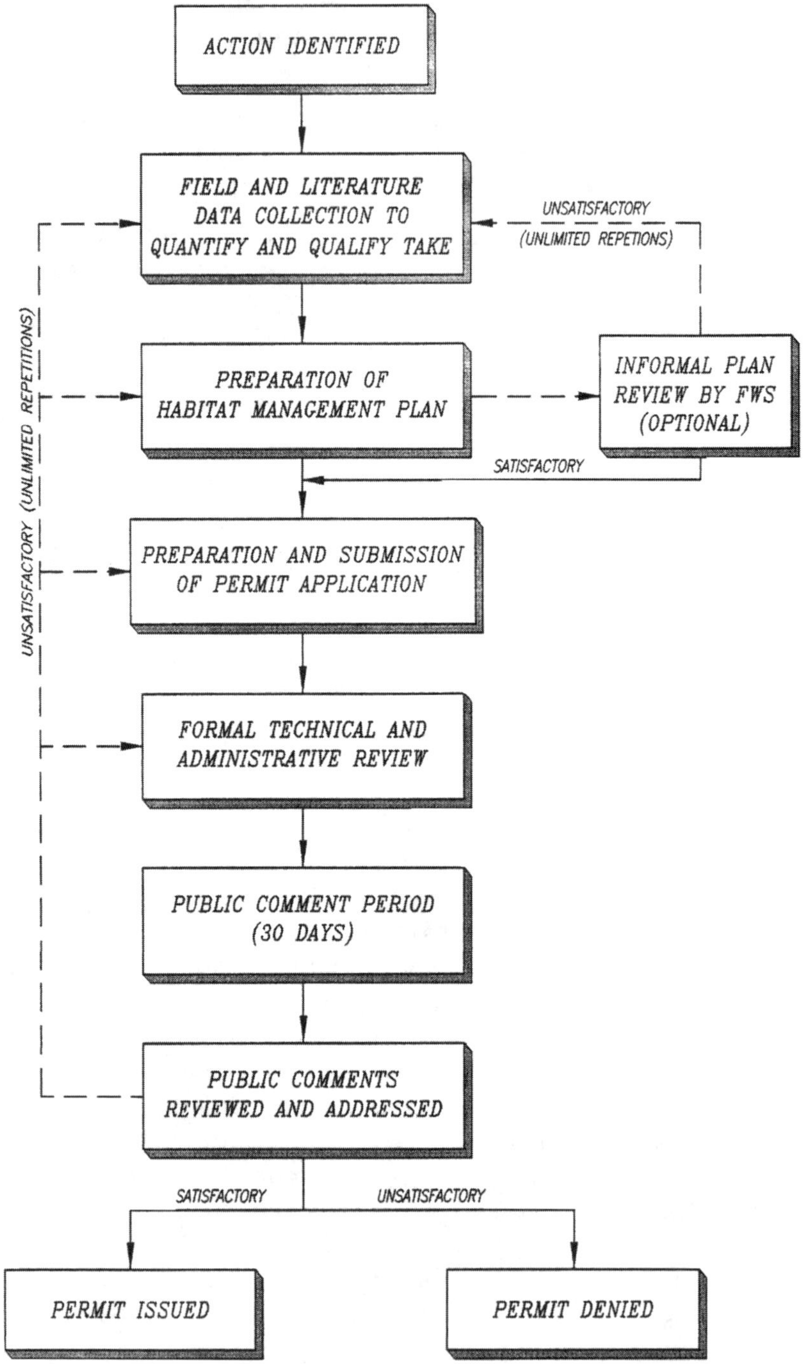

the implementation of the habitat management plan may also cause the project to appear too expensive to pursue. Incidental taking permits have their greatest utility in situations where the exploration and production project has a high likelihood of being economically viable and the production is anticipated to continue for an extended time period.

ESA *Section* 11—*Penalties and* Enforcement

The ESA provides for both civil and criminal penalties. Civil penalties can be assessed in amounts of up to $25,000 per violation. For taking violations, each individual of a species that is taken can be regarded as a separate violation. Criminal violations of the ESA are punishable by fines of up to $50,000 per violation and imprisonment for up to one year. The knowing violation of permit conditions of a Section 7 taking statement limitation can be subject to criminal penalties under the ESA.

Only one defense is specifically provided for in the ESA.

> *. . . if it can be shown by a preponderance of the evidence that the defendant committed an act based on a good faith belief that he was acting to protect himself or herself, a member of his or her family, or any other individual from bodily harm from any threatened or endangered species*[61]

then no civil or criminal penalties will be imposed. This defense has proved of little practical significance. The courts have found their credulity strained by individuals attempting to show, by the preponderance of the evidence, that they were defending themselves against such species as the jaguar, grizzly bear, or Florida panther. It seems likely that they would be even more incredulous if such arguments were applied to species such as the noonday snail, mission blue butterfly, golden-cheeked warbler, or the Houston toad.

Current Controversy over the ESA

The ESA has become, of late, one of the most controversial of our environmental statutes. This is due, primarily, to the regulatory definition of harm. As was previously discussed, the regulatory definition of harm includes habitat modification that actually kills or injures wildlife by significantly impairing essential behavioral patterns. This definition is being challenged before the Supreme Court at the time of this writing.[62] It is being argued that it was the intent of Congress that harm mean actual physical harm to the individuals of the species at the time of the action. This line of argument would appear to lead to a nonsensical outcome since, if literally applied, such a definition could conceivably allow for the decimation of the entire breeding range of a migratory species such as Kirtland's warbler (*Dendroica kirtlandii*) while it is on the winter range, since no individuals would be present to be directly harmed.

The function of the ESA is also being challenged by draft amendments of the act that would require compensation for a regulatory taking, if the protection of a listed species resulted in some threshold diminution of property values. This would probably be of limited significance in exploration and production since many projects require federal actions and ESA Section 7's mandate that agencies use their authorities to preserve listed species has not been challenged.

APPENDIX 10 ENDNOTES

1. 16 USC §§ 1531 to 1544.
2. Gunn, Alastair S.: "Why Should We Care About Rare Species?" *Environmental Ethics,* Volume 2 Number 2 (Spring 1980), p. 19.
3. 16 USC § 1531(a)(1), (2), (3).
4. U.S. Department of the Interior, Fish and Wildlife Service. "Box Score", *Endangered Species Bulletin,* Volume *XX* Number 1 (January/February 1995), p. 24.
5. 16 USC § 1531(6).
6. 16 USC § 1531(c).
7. 16 USC § 1538.
8. 16 USC § 1536.
9. 16 USC § 1538.
10. 16 USC § 1539.
11. 16 USC § 1532(16).
12. 16 USC § 1532(6).
13. 16 USC § 1532(20).
14. 16 USC § 1532(6).
15. 16 USC § 1533(a)(1).
16. 50 CFR § 424.11(b) (1994).
17. H.R. Conf. Rep. No. 835, 97th Cong., 2d Sess. 21, reprinted in 1982 U.S. Code Cong. and Admin. News at 2861.
18. 16 USC § 1533(b)(3)(B)(iii).
19. Id.
20. 16 USC § 1533(b)(7).
21. 16 USC § 1533(b)(3)(B)(i).
22. 16 USC § 1533(b)(3)(B)(ii).

23. 16 USC § 1533(b)(3)(B)(iii).
24. 16 USC § 1533(b)(37)(C)(i).
25. 16 USC § 1533(b)(3)(B)(iii).
26. 50 CFR § 424.16(C)(2) (1994).
27. 50 CFR §§ 424.17(a)(1)(i), (iii), (iv) (1994).
28. 16 USC § 1533(b)(7); 50 CFR § 424.20 (1994).
29. 16 USC § 1533(b)(2).
30. Quarles, William R.: "Construction on all spreads either complete or under way: Kern River Mojave Update", *Pipe Line Industry* (1991).
31. 16 USC § 1536(a)(2).
32. 50 CFR § 402.08 (1994).
33. 50 CFR § 402.12(c) (1994).
34. 50 CFR § 402.12(d)(1) (1994).
35. 50 CFR § 402.13(a) (1994).
36. 50 CFR § 402.13(b) (1994).
37. 50 CFR § 402.12 (1994)
38. 50 CFR § 402.12(f) (1994).
39. 50 CFR § 402.12(e) (1994).
40. 50 CFR § 402.12(i) (1994).
41. 50 CFR § 402.12(a) (1994).
42. 50 CFR § 402.12(j) (1994).
43. 50 CFR § 402.14(a) (1994).
44. 50 CFR § 402.14(c) (1994).
45. 50 CFR § 402.14(d) (1994).
46. 50 CFR § 402.14(e) (1994).
47. Id.
48. 50 CFR § 402.114(i) (1994).
49. 50 CFR § 402.14(i)(2) (1994).
50. 50 CFR § 402.15(b) (1994).
51. 16 USC § 1532(9).
52. 5 CFR § 17.3 (1994).
53. Id.
54. Id.
55. Thornton, Robert D.: "Takings Under Endangered Species Act Section 9," *Natural Resources and Environment,* volume 4 number 4 (1990), p. 7.
56. 16 USC § 1539(a)(1)(B).
57. 16 USC § 1539(a)(2)(A).
58. 16 USC § 1539(a)(2)(B).
59. 16 USC § 1539(a)(2)(C).
60. 50 CFR §§ 17.22 and 17.32.
61. 16 USC §§ 1540(a)(3) and (b)(3).
62. Sweet Home Chapter of Communities for a Greater Oregon v. Babbitt 1 F.3d 1(D.C. Cir. 1993) and 17 F.3d 1463 (D.C. Cir. 1994).

INDEX